Julia Hörrmann

THE METHOD OF DENSITIES FOR NON-ISOTROPIC BOOLEAN MODELS

The method of densities for non-isotropic Boolean models

by

Julia Hörrmann

KIT Scientific Publishing

Dissertation, Karlsruher Institut für Technologie (KIT)
Fakultät für Mathematik, 2014
Tag der mündlichen Prüfung: 22. Oktober 2014

Impressum

KIT Scientific Publishing

Karlsruher Institut für Technologie (KIT)
KIT Scientific Publishing
Straße am Forum 2
D-76131 Karlsruhe

KIT Scientific Publishing is a registered trademark of Karlsruhe Institute of Technology. Reprint using the book cover is not allowed.

www.ksp.kit.edu

Print on Demand 2015

ISBN 978-3-7315-0329-3
DOI 10.5445/KSP/1000045101

The method of densities for non-isotropic Boolean models

Zur Erlangung des akademischen Grades eines
DOKTORS DER NATURWISSENSCHAFTEN

von der Fakultät für Mathematik des
Karlsruher Instituts für Technologie (KIT)

genehmigte
DISSERTATION

von

Dipl.-Math. techn. Julia Hörrmann
aus Karlsruhe

Tag der mündlichen Prüfung:	22. Oktober 2014
Referent:	Prof. Dr. Daniel Hug
Korreferenten:	Prof. Dr. Eva B. Vedel Jensen
	Prof. Dr. Wolfgang Weil

Abstract

The Boolean model is the basic random set model for the description of porous structures like the pore space in sandstone or bread or for the solid phase of sintered ceramic composites. It is obtained by decorating the points of a homogeneous Poisson point process (the germs) with independent identically distributed random compact, convex particles (the grains) and forming the union set. If a Boolean model is fitted to a real structure the first task is to estimate the intensity from observable quantities.

In the isotropic case the Miles formulas can be used to express the intensity in terms of the densities of the intrinsic volumes (in two dimensions these are the volume, the half of the surface area and the Euler characteristic). This approach is known as the method of densities.

In the more difficult non-isotropic situation Weil showed in two and three dimensions that the intensity is uniquely determined by the densities of the mixed volumes.

Combining Weil's ideas of translative integral geometry with ideas from harmonic analysis and approximation theory we obtain formulas for the intensity, which are directly comparable to the Miles formulas.

For this purpose we introduce a new collection of geometric functionals on the space of convex bodies, the *harmonic intrinsic volumes*.

Under a regularity assumption on the grain distribution we obtain a series representation of the intensity in terms of the densities of the harmonic intrinsic volumes.

Moreover, we introduce a *modulus of isotropy* of the grain distribution to quantify the degree of anisotropy of the Boolean model. If the intensity is approximated by the truncated series depending on the densities of only finitely many harmonic intrinsic volumes, we can bound the error term from above using the modulus of isotropy.

Acknowledgement

This PhD thesis has been written during my employment as research and teaching assistant at the Institute of Stochastics at the Karlsruhe Institute of Technology (KIT). My position was financially supported by the German Research Foundation (DFG) through the research unit "Geometry and Physics of Spatial Random Systems" under the grant HU 1874/3-1. In this place I wish to thank several people.
Special thanks go to my advisor Prof. Dr. Daniel Hug for introducing me to a great variety of topics in Stochastic Geometry, for valuable comments through the last years and for giving me a lot of freedom. My sincere thanks go to my reviewers Prof. Dr. Eva B. Vedel Jensen and Prof. Dr. Wolfgang Weil. I also wish to thank Prof. Dr. Günter Last and Prof. Dr. Wolfgang Weil for some very helpful remarks concerning especially the disintegration results of this thesis.
Thanks go also to my collaborators Michael Klatt, Prof. Dr. Klaus Mecke, Prof. Dr. Matthias Reitzner and Prof. Dr. Christoph Thäle. I learned a lot from all of them. I would also like to thank Prof. Dr. Markus Kiderlen for inspiring discussions during my stay at the Centre for Stochastic Geometry and Advanced Bioimaging (CSGB) at Aarhus University.
I owe huge thanks to Dr. Matthias Schulte for his valuable comments and his support in the final months. I would like to thank all members of the Institute of Stochastics for creating a friendly working atmosphere. Finally, I am deeply indebted to my family and to Michael Wolf for their encouragement in the last years.

Contents

1 Introduction **1**

2 Foundations **9**
2.1 Foundations from probability theory 9
2.2 Foundations from convex geometry 10
2.3 Foundations from stochastic geometry 12
2.4 Spherical harmonics and two auxiliary lemmas 15
2.5 A decomposition of the Haar measure on the rotation group 19

3 Translative integral geometric results for support and area measures **21**
3.1 Support and area measures 21
3.2 Translative integral formulas 23
3.2.1 The translative integral formula 23
3.2.2 Properties of mixed support measures 24
3.3 Partial inversion of the translative integral formula 30
3.4 Representation of special mixed support measures 32

4 Density formulas for area measures of Boolean models **37**
4.1 Mean value and density formulas 37
4.1.1 The isotropic situation 37
4.1.2 The non-isotropic situation 38
4.1.3 Density formulas for mixed area measures 42
4.2 Inversion of density formulas for area measures 42

5 Application to Minkowski tensors **47**
5.1 Minkowski tensors 47
5.2 Mixed Minkowski tensors 50
5.3 Mean value and density formulas for Minkowski tensor 51
5.3.1 The isotropic situation 51
5.3.2 The non-isotropic situation 52
5.3.3 Density formulas for mixed Minkowski tensors 53
5.4 Application to a parametric Boolean model in the plane 54
5.5 Planar Boolean model with smooth grains 59
5.6 Uniqueness results for Minkowski tensors 60

6 Application to harmonic intrinsic volumes **63**
6.1 Harmonic intrinsic volumes 63
6.2 Mixed harmonic intrinsic volumes 65
6.3 Mean value and density formulas for harmonic intrinsic volumes . . 67
6.3.1 The isotropic situation 67
6.3.2 The non-isotropic situation 67
6.3.3 Density formulas for mixed harmonic intrinsic volumes . . . 68
6.4 Uniqueness results for harmonic intrinsic volumes 69

7 The method of densities via harmonic intrinsic volumes **71**
7.1 A disintegration result 72
7.2 Rotation regularity 78
7.3 Absolute continuity of densities of area measures 82
7.4 A modulus of isotropy for rotation regular particle processes 87
7.4.1 Motivation 88
7.4.2 A modulus of smoothness on the unit sphere 89
7.4.3 A modulus of smoothness on the rotation group 92
7.4.4 A modulus of isotropy for a rotation regular particle process . 96
7.5 Polynomial approximation of Radon-Nikodym derivatives 98
7.6 A lower bound for the modulus of isotropy 103
7.7 Inversion of density formulas for harmonic intrinsic volumes 105
7.7.1 The two dimensional case 106
7.7.2 The three dimensional case 108

1 Introduction

Introduced in 1975 by Matheron [Mat75] the Boolean model is now the basic random closed set model for the description of porous structures. It is obtained by forming the union set of a Poisson particle process on the space of compact convex sets. Depending on the specific application, either the pore space or the solid phase of a material may be described as a Boolean model. For example, the pore space of bread [BS91] was modeled by the Boolean model, whereas for sintered ceramic composites it is the solid phase which is described as a Boolean model; see [RG00]. In particular physical phenomena like percolation [MW91, MS02] and elasticity [AKM09] can be studied using Boolean models. A treatment of the Boolean model, which contains a collection of its various applications, can be found in [CSKM13, Chapter 3]. For an introduction of the Boolean model, which is accessible to practitioners and presents available statistical methods, we refer to the monograph [Mol97].

Let $\{\xi_i : i \in \mathbb{N}\}$ be a random collection of points in $\mathbb{R}^d$ forming a stationary Poisson point process with intensity $\gamma > 0$. Let $Z_0, Z_1, Z_2, \dots$ be independent, identically distributed random convex bodies (nonempty, compact, convex sets) with distribution $\mathbb{Q}$, which are independent of the point process $\{\xi_i : i \in \mathbb{N}\}$. The random points $\xi_1, \xi_2, \dots$ are the germs and the random sets $Z_1, Z_2, \dots$ are the grains of the Boolean model. The random set Z_0 is called the typical grain. We can assume without loss of generality that the centre of the circumball of the typical grain is almost surely at the origin. Then, under a mild integrability condition on the grain distribution, the union of the translated grains

$$Z := \bigcup_{i=1}^{\infty} (Z_i + \xi_i)$$

is a random closed set, which is called the stationary Boolean model with intensity γ and grain distribution $\mathbb{Q}$. The random collection $X := \{\xi_1 + Z_1, \xi_2 + Z_2, \dots\}$ of the shifted grains is the particle process underlying the Boolean model. Alternatively the Boolean model can be introduced directly as the union set of a particle process X on the space of compact, convex sets, see Section 2.3 for this approach.

Assume we observe Z in a compact, convex observation window W with positive volume. Our aim is to extract distributional information from the geometric properties of the sample $Z \cap W$. Thus, we assume we can measure geometric functionals like the volume or the surface area of the sample. The sample $Z \cap W$ is a finite union of convex bodies (a polyconvex set). By a geometric functional φ we mean a real-valued functional defined on the space of polyconvex sets with special additional properties, see the precise definition in Section 2.3. Important examples of geometric functionals are the intrinsic volumes $V_d, \dots, V_0$, which are the coefficients in the polynomial expansion for the volume of the parallel set of a convex body. For polyconvex sets which are the closure of their interior $2V_{d-1}$ is the surface area. The intrinsic volume V_0 is the Euler characteristic, which is constantly equal to 1 on the space of convex bodies. In two dimensions V_0 is for polyconvex sets equal to the number of connected components minus the number of holes. The boundary of the observation window W has a disturbing effect. It is therefore of advantage to assume a sufficiently large observation window and to consider only limits as the window tends to infinity. This motivates the introduction of the density (or specific value) of the Boolean model for a geometric functional φ. The density of φ is the combined spatial and probabilistic mean value

$$\overline{\varphi}(Z) = \lim_{r \to \infty} \frac{\mathbb{E}\varphi(Z \cap rW)}{V_d(rW)}.$$

The crucial problem when studying a Boolean model is, that the particles overlap and can therefore not be observed individually. Thus, neither the intensity nor the mean geometric properties of the typical grain can be estimated directly from observations. On the other hand it can be shown that the mean value $\mathbb{E}\varphi(Z \cap W)$ of a geometric functional of a sample has a representation as an alternating series of mean values of so-called translative integrals

$$\mathbb{E} \int_{(\mathbb{R}^d)^k} \varphi(W \cap (Z_1 + x_1) \cap \dots \quad \dots \cap (Z_k + x_k)) d(x_1, \dots, x_k), \qquad k \in \mathbb{N}.$$

Thus, to extract distributional properties from the above mean value or from the density $\overline{\varphi}(Z)$, it is necessary to study translative integrals of geometric functionals. This is the topic of translative integral geometry, see [SW08, Sections 5.2 and 6.4]. In the ideal case translative geometric results can be used to replace the application of φ to the intersection of translates of $Z_1, \dots, Z_k$ with the observation window by an expression involving only geometric functionals of the individual grains $Z_1, \dots, Z_k$. The most basic results of this form can be obtained if φ is the volume V_d or V_{d-1}, which is half of the surface area. Then, one can obtain the density formulas

$$\overline{V}_d(Z) = 1 - e^{-\gamma \mathbb{E} V_d(Z_0)} \text{ and } \overline{V}_{d-1}(Z) = e^{-\gamma \mathbb{E} V_d(Z_0)} \gamma \mathbb{E} V_{d-1}(Z_0). \tag{1.1}$$

Here, mean values of the volume respectively the surface area of the typical grain multiplied by the intensity γ occur on the right-hand side. It is therefore convenient to introduce for a geometric functional φ the notation

$$\overline{\varphi}(X) := \gamma \mathbb{E} \varphi(Z_0).$$

The value $\overline{\varphi}(X)$ is called the density of the particle process X since it has a representation similar to the definition of $\overline{\varphi}(Z)$ where the geometric functional φ is applied to the grains individually instead of to the union set Z and the contributions of the grains are summed up, compare (2.5). Thus relation (1.1) reads now

$$\overline{V}_d(Z) = 1 - e^{-\overline{V}_d(X)} \text{ and } \overline{V}_{d-1}(Z) = e^{-\overline{V}_d(X)}\overline{V}_{d-1}(X). \tag{1.2}$$

The density formulas from (1.2) can be inverted, which yields the formulas

$$\overline{V}_d(X) = -\ln(1 - \overline{V}(Z)) \text{ and } \overline{V}_{d-1}(X) = \frac{1}{1 - \overline{V}_d(Z)}\overline{V}_{d-1}(Z). \tag{1.3}$$

In (1.3) the densities of the particle process X are expressed by the observable densities of the Boolean model Z. However, the actual aim is to extract from observations of the Boolean model information about its parameters γ and $\mathbb{Q}$ separately. That is, we would like either to extract the intensity γ or mean values of geometric functionals of the typical grain.

The first task when fitting a Boolean model to a real structure is therefore the estimation of the intensity γ – on the one hand as a fundamental parameter on its own, on the other hand because it is needed if one wants to extract mean values of geometric functionals of the typical grain from the densities of the underlying particle process. For an overview of the different existing methods for the estimation of the intensity we refer to [Mol97, Chapter 5], [SW08, Chapter 9.5] and [CSKM13, Chapter 3.4.3].

One of the most popular methods seems to be the method of densities. Its idea is to proceed similarly as for the method of moments in statistics where model parameters are chosen in such a way that the moments of a parametric distribution on the real line coincide with empirical moments. The role of the empirical moments is now played by estimators of the densities of the Boolean model whereas the model parameters we are interested in are the intensity and the mean values of geometric functionals of the typical grain.

However, the state of the art is that the idea of the method of moments can be directly applied only for the isotropic Boolean model, that is for Boolean models Z with a distribution which is invariant under rotations of Z. The reason is that only for an isotropic Boolean model density formulas as in (1.2) can be obtained for all intrinsic volumes. These formulas are known as the Miles formulas, see Miles [Mil76] and Davy [Dav76]. They express the observable densities $\overline{V}_d(Z), \ldots, \overline{V}_0(Z)$ in terms of the densities $\overline{V}_d(X), \ldots, \overline{V}_0(X)$ and can be inverted. For example in two dimensions a complete set of density formulas for the intrinsic volumes is given by

$$\begin{aligned} \overline{V}_2(Z) &= 1 - e^{-\overline{V}_2(X)}, \\ \overline{V}_1(Z) &= e^{-\overline{V}_2(X)}\overline{V}_1(X), \end{aligned} \tag{1.4}$$

$$\overline{V}_0(Z) = e^{-\overline{V}_2(X)} \left[\overline{V}_0(X) - \frac{1}{\pi} \overline{V}_1(X)^2 \right].$$

Their practical relevance relies on the relation $\overline{V}_0(X) = \gamma$, which implies that the Miles formulas can be used to determine from the densities $\overline{V}_d(Z), \ldots, \overline{V}_0(Z)$ of the Boolean model the intensity γ and all mean values $\mathbb{E}V_d(Z_0), \ldots, \mathbb{E}V_0(Z_0)$. For example we have in two dimensions

$$\gamma = \rho \overline{V}_0(Z) + \rho^2 \frac{1}{\pi} \overline{V}_1(Z)^2, \tag{1.5}$$

where we use the abbreviation $\rho := (1 - \overline{V}_2(Z))^{-1}$. This is not only a surprising fact, it is also very useful for applications and the basis for the application of the method of densities for an isotropic Boolean model.

For a non-isotropic Boolean model the situation is much more difficult because the density formulas for the intrinsic volumes $V_{d-2}, \ldots, V_0$ involve so-called mixed functionals depending on several convex bodies. For example in two dimensions a functional $V_{1,1}$ depending on two convex bodies occurs. Then, the density formula for the Euler characteristic is

$$\overline{V}_0(Z) = e^{-\overline{V}_2(X)} \left[\overline{V}_0(X) - \overline{V}_{1,1}(X, X) \right]$$

with the mixed density

$$\overline{V}_{1,1}(X, X) := \gamma^2 \mathbb{E} V_{1,1}(Z_0, Z_1).$$

In higher dimensions the formulas become more and more difficult. In three dimensions the density formulas for V_1 involve the mixed functional $V_{2,2}$, which depends on two convex bodies, and the density formula for V_0 involves the mixed functionals $V_{2,1}$, which depends on two, and $V_{2,2,2}$, which depends even on three convex bodies. A simple inversion formula for the intensity as (1.5) is therefore out of reach.

The investigation of the mixed functionals and the search for new geometric functionals which complement the set of intrinsic volumes in an elegant way was mainly promoted by Wolfgang Weil, see for example his lecture notes [Wei07] and the references therein. In the non-isotropic case the method of densities is therefore also called Weil's method. However, this line of research cannot be considered as completed because so far only uniqueness results are available. In [Wei99] it is shown that the intensity in two and three dimensions is uniquely determined by the densities of the intrinsic volumes and the densities of more complicated geometric objects like the centred support function and the surface area measure, which can be associated to a polyconvex set. A later uniqueness result in [Wei01a] involves only real-valued geometric functionals. Namely, the densities of the mixed volumes V, which depend in d dimensions on d polyconvex sets, are needed. More precisely, it is shown that in two dimensions the densities of the intrinsic volumes together with the collection of densities of the geometric functionals $K \mapsto V(K, M)$ where M is a fixed arbitrary convex body determine the intensity, whereas in three dimensions the densities of the intrinsic volumes together with the collection of densities of the geometric functionals $K \mapsto V(K, M, M)$ and $K \mapsto V(K, K, M)$ where M is a fixed arbitrary convex body determine the intensity.

With the aim to contribute to the method of densities and in particular to the important problem of intensity estimation for non-isotropic Boolean models we introduce in this thesis a new collection of geometric functionals, which complements the intrinsic volumes in a natural way, the *harmonic intrinsic volumes*. They are obtained as integrals of the area measures $\Psi_{d-1}, \dots, \Psi_0$, which are measures on the unit sphere and can be considered as local versions of the intrinsic volumes. The integrands are orthonormal homogeneous polynomials on the unit sphere. More precisely, the harmonic intrinsic volume $V_j^{l,p}$ is a geometric functional obtained as the integral with respect to the area measure Ψ_j of a homogeneous polynomial of polynomial degree l on the unit sphere. The second exponent p is used to enumerate an orthonormal basis of the homogeneous polynomials of polynomial degree l. An appealing property of the harmonic intrinsic volumes is the rotation formula

$$\int_{SO_d} V_j^{l,p}(\vartheta K)\nu(d\vartheta) = 0, \quad l > 0,$$

for polyconvex sets K, where SO_d is the group of proper rotations and ν the Haar probability measure on this group. As a main result we obtain in the final section of this thesis inversion formulas for the intensity in two and three dimensions under a mild regularity condition on the grain distribution.

Theorem 1. *In two dimensions the intensity γ has the series representation*

$$\gamma = \rho \overline{V}_0(Z) + \rho^2 \left[\frac{1}{\pi} \overline{V}_1(Z)^2 + \sum c_{l,m}^{p,q} \, \overline{V}_1^{l,p}(Z) \overline{V}_1^{m,q}(Z) \right]$$

with constants $c_{l,m}^{p,q}$ for $l + m > 0$ and $\rho := (1 - \overline{V}_2(Z))^{-1}$.

For an isotropic Boolean model the sum over the densities of the harmonic intrinsic volumes vanishes, which yields the well-known result (1.5), which can be obtained by using the Miles formulas. In this thesis we introduce also a modulus of isotropy of the underlying particle process as a measure of the degree of anisotropy of the Boolean model. In the final section we obtain upper bounds for the error if the intensity is approximated by the truncated sum involving only finitely many densities of harmonic intrinsic volumes.

As a corresponding result in three dimensions we obtain the following theorem.

Theorem 2. *In three dimensions the intensity γ has the series representation*

$$\gamma = \rho \overline{V}_0(Z) + \rho^2 \left[d\overline{V}_1(Z)\overline{V}_2(Z) + \sum d_{l,m}^{p,q} \overline{V}_1^{l,p}(Z)\overline{V}_2^{m,q}(Z) \right] + \rho^3 \left[e\overline{V}_2(Z)^3 + \sum e_{l,m,o}^{p,q,s} \overline{V}_2^{l,p}(Z)\overline{V}_2^{m,q}(Z)\overline{V}_2^{o,s}(Z) \right]$$

with constants $d, d_{l,m}^{p,q}$ for $l + m > 0$ and constants $e, e_{l,m,o}^{p,q,s}$ for $l + m + o > 0$ and $\rho := (1 - \overline{V}_3(Z))^{-1}$.

Again all sums over densities of harmonic intrinsic volumes vanish for an isotropic Boolean model.

An open question remains the derivation of corresponding formulas for the intensity in dimensions higher than three.

In this thesis we restricted to stationary Boolean models with convex grains. Under an additional integrability assumption on the grain distribution it should be possible to transfer our results to Boolean models with polyconvex grains whose Euler characteristic is equal to one, compare [Wei01a] for the principle approach.

The structure of the thesis is the following. Chapter 2 contains the necessary background material from probability theory, convex geometry and stochastic geometry, some information on spherical harmonics and some auxiliary results.

The starting point of this thesis has been the idea to apply the translative integral geometric results for support measures which were obtained by Daniel Hug in his habilitation thesis [Hug99]. These results are recalled in Chapter 3 and complemented by new representations for mixed support and area measures in the concluding two sections of Chapter 3.

In Chapter 4 Daniel Hug's results are applied to obtain density formulas for the translation invariant marginal measures of the support measures, the area measures. In the final section of Chapter 4 we apply the new representations for mixed measures from the previous chapter to obtain inversion results for the densities of the area measures. In particular we obtain in the final section integral representations of the intensity of a non-isotropic Boolean model in two and three dimensions.

Chapter 5 is indebted to the original aim of this thesis, namely the application of Daniel Hug's translative integral geometric results for the derivation of density formulas of the Minkowski tensors. These tensor valued geometric functionals have various physical applications and are also interesting from a purely convex geometric viewpoint. In this chapter we consider also a parametric example of a Boolean model and obtain uniqueness results for the Minkowski tensors, which are in the spirit of Weil's results for mixed volumes mentioned above.

In Chapter 6 we introduce the new harmonic intrinsic volumes and derive density formulas and uniqueness results similar to the results for the Minkowski tensors in the previous chapter.

The final Chapter 7 contains the main contributions to the method of densities for non-isotropic Boolean models.

It is a collection of various different results, which are combined in the final section. In the first section, Section 7.1, we obtain a disintegration result for the grain distribution, which is used in Section 7.2 for the introduction of the property of rotation regularity. This regularity assumption has the pleasant effect that the densities of the area measures are absolutely continuous with respect to the spherical Lebesgue measure as we show in Section 7.3. Motivated by the concept of a modulus of continuity we introduce in Section 7.4 the notion of a modulus of isotropy for a rotation regular particle process. In Section 7.5 we derive upper bounds for the best polynomial approximation of the Radon-Nikodym derivatives of the densities of the area measures with respect to the L^2-norm. These bounds depend on the modulus of isotropy. In Section 7.6 we relate the modulus of isotropy to the Euclidean norm of the vector of densities of the harmonic intrinsic volumes of fixed polynomial degree. The main results are contained in the concluding Section 7.7. Here also Theorem 1 and Theorem 2, which have been presented in the introduction, are achieved in Theorem 7.7.1 respectively the second part of Theorem 7.7.6. In two dimensions we obtain in Theorem 7.7.1 a series representation of the intensity in terms of densities of the harmonic intrinsic volumes. In Theorem 7.7.2 we obtain error bounds in terms of the modulus of isotropy of the underlying particle process if the intensity is approximated by the truncated series. Theorem 7.7.6 contains in three dimensions a series representation for the intensity in terms of the densities of the harmonic intrinsic volumes and also a corresponding representation for the densities of the particle process of the harmonic intrinsic volumes $V_1^{l,p}$. In Theorem 7.7.8 and Theorem 7.7.9 we obtain bounds for the approximation error in terms of the modulus of isotropy of the underlying particle process if only truncated series are used for the approximation of the intensity and the density of the harmonic intrinsic volume $V_j^{l,p}$, respectively.

2 Foundations

2.1 Foundations from probability theory

We denote the underlying probability space by $(\Omega, \mathcal{A}, \mathbb{P})$. The distribution of a random variable ξ is denoted by $\mathbb{P}_\xi$. If ξ is a real random variable, we denote by $\mathbb{E}\,\xi$ its expected value. A measure or signed measure on a topological space E will be defined on the σ-algebra $\mathcal{B}(E)$ of Borel sets if not stated otherwise. A finite signed measure μ on E can be represented as the difference of its positive part

$$\mu^+(A) := \sup\{\mu(B) : B \in \mathcal{B}(E) \text{ and } B \subset A\}, \quad A \in \mathcal{B}(E),$$

and its negative part

$$\mu^-(A) := \sup\{-\mu(B) : B \in \mathcal{B}(E) \text{ and } B \subset A\}, \quad A \in \mathcal{B}(E).$$

The representation $\mu = \mu^+ - \mu^-$ of μ as difference of the two positive measures μ^+ and μ^- is called Jordan decomposition, see [Coh13, p. 117]. We denote by $C(E)$ the space of all real-valued continuous functions on E and by $C_b(E)$ the space of functions in $C(E)$ which are bounded. Let $\mu, \mu_i, i \in \mathbb{N}$, be finite positive measures on E. The sequence $(\mu_n)_{n\in\mathbb{N}}$ converges weakly to μ if

$$\lim_{n\to\infty} \int_E f(e)\mu_n(de) = \int_E f(e)\mu(de), \quad f \in C_b(E).$$

For two topological spaces E and F, a mapping $\mu : E \times \mathcal{B}(F) \to [0, \infty]$ is called a probability kernel from E to F if $\mu(\cdot, B)$ is Borel measurable for fixed $B \in \mathcal{B}(F)$ and if $\mu(e, \cdot)$ is a probability measure for fixed $e \in E$. For $e \in E$ the Dirac measure δ_e is defined by $\delta_e(A) := \mathbf{1}\{e \in A\}$ for $A \in \mathcal{B}(E)$. The Lebesgue measure on $\mathbb{R}^d$ is denoted by λ. The k-dimensional Hausdorff measure is denoted by $\mathcal{H}^k$. By $\mathcal{H}^k \llcorner A$ we denote the restriction of $\mathcal{H}^k$ to a subset A.

For $p \in [1, \infty]$ and a measure space $(E, \mathcal{B}(E), \mu)$ we use the notation $L^p(E, \mu)$ for the L^p-space and $\|\cdot\|_p$ for the L^p-norm. We extend the definition of $\|\cdot\|_p$ to an arbitrary real-valued measurable function f on E, which means that $\|f\|_p$ may be equal to infinity.

For $f, g \in L^2(E, \mu)$ we denote the scalar product by

$$(f, g) := \int_E f(e) g(e) \mu(de).$$

If we identify functions which coincide μ-almost everywhere, the space $L^2(E, \mu)$ is a Hilbert space with scalar product $(\cdot, \cdot)$.

2.2 Foundations from convex geometry

For notions from convex geometry we refer to [SW08, Chapter 14] and [Sch13a]. Let $\langle \cdot, \cdot \rangle$ be the scalar product and $\| \cdot \|$ the norm in $\mathbb{R}^d$. Let $e_1, \ldots, e_d$ denote the standard orthonormal basis of $\mathbb{R}^d$. The system of nonempty, compact subsets of $\mathbb{R}^d$ is denoted by $\mathcal{C}$. On $\mathcal{C}$ we use the Hausdorff metric δ, which is defined by

$$\delta(K, L) := \max \left\{ \max_{x \in K} \min_{y \in L} \|x - y\|, \max_{x \in L} \min_{y \in K} \|x - y\| \right\}, \quad K, L \in \mathcal{C}.$$

By $\mathcal{K}$ we denote the family of nonempty, compact, convex subsets (convex bodies) of $\mathbb{R}^d$.

The convex ring $\mathcal{R}$ consists of all finite unions of convex bodies and its elements are called polyconvex sets. We also use the Hausdorff metric on the subclasses $\mathcal{K}$ and $\mathcal{R}$ of $\mathcal{C}$.

The extended convex ring $\mathcal{S}$ is the system of sets whose intersection with any compact convex set belongs to the convex ring and its elements are called locally polyconvex sets. Let $\mathcal{M}$ be a family of sets which is closed under intersections. A functional φ on $\mathcal{M}$ with values in an abelian group G is called additive (or a valuation) if

$$\varphi(K \cup L) = \varphi(K) + \varphi(L) - \varphi(K \cap L), \quad K, L \in \mathcal{M} \text{ with } K \cup L \in \mathcal{M}.$$

If $\emptyset \notin \mathcal{M}$ we extend the definition of φ by $\varphi(\emptyset) = 0$. The most important cases in this thesis are $\mathcal{M} = \mathcal{K}$ and $\mathcal{M} = \mathcal{R}$.

Let A be a subset of $\mathbb{R}^d$. Then $\operatorname{int} A, \partial A$ and $\operatorname{relint} A$ are, respectively, the interior, the boundary and the relative interior of A. For $A, B \subset \mathbb{R}^d$ we use the notation

$$A + B := \{a + b : a \in A, b \in B\}$$

for the Minkowski addition of the sets A and B. We denote the unit ball by B^d, the unit sphere by S^{d-1} and the unit cube by $C^d := [0, 1]^d$. The volume of the unit ball B^d is given by $\kappa_d := \lambda(B^d) = \pi^{d/2}/\Gamma(1+d/2)$ and the surface area of the unit sphere S^{d-1} by $\omega_d := d\kappa_d$. For $K \in \mathcal{K}$ the Steiner formula

$$\lambda_d(K + \epsilon B^d) = \sum_{j=0}^{d} \epsilon^{d-j} \kappa_{d-j} V_j(K), \quad \epsilon > 0, \tag{2.1}$$

defines the intrinsic volumes $V_0, \dots, V_d$. On $\mathcal{K}$ the intrinsic volumes are nonnegative, monotone under set inclusion, additive, continuous and invariant under rotations and translations. They have a unique additive extension to $\mathcal{R}$. On $\mathcal{R}$ the functional V_d is the Lebesgue measure. If $K \in \mathcal{R}$ is the closure of its interior, $V_{d-1}(K)$ is the half of the surface area of K. The functional V_0 is the Euler characteristic, that is the additive functional with $V_0(K) = 1$ for $K \in \mathcal{K}$. For $m \in \mathbb{N}$ and $K_1, \dots, K_m \in \mathcal{K}$ the polynomial expansion

$$V_d(\lambda_1 K_1 + \ldots + \lambda_m K_m) = \sum_{i_1,\dots,i_d=1}^{m} \lambda_{i_1} \cdots \lambda_{i_d} V(K_{i_1}, \dots, K_{i_d}), \quad \lambda_1, \dots, \lambda_m > 0, \tag{2.2}$$

defines the mixed volume $V : (\mathcal{K})^d \to \mathbb{R}$. The mixed volume V is additive with respect to each argument. For $A \subset \mathbb{R}^d$ we denote by

$$\operatorname{pos} A := \left\{ \sum_{i=1}^{k} \lambda_i x_i : \lambda_i \geq 0, x_i \in A, 1 \leq i \leq k, k \in \mathbb{N} \right\}$$

the positive hull of A. For $u_1, \dots, u_k \in S^{d-1}$ let $\nabla_k(u_1, \dots, u_k)$ be the volume of the parallelepiped spanned by $u_1, \dots, u_k$ and

$$\operatorname{sconv}(u_1, \dots, u_k) := \operatorname{pos}\{u_1, \dots, u_k\} \cap S^{d-1}$$

the spherical convex hull of $u_1, \dots, u_k$. The spherical Lebesgue measure on S^{d-1} is denoted by σ and scaled in such a way that $\sigma\left(S^{d-1}\right) = \omega_d$.

A convex body $K \in \mathcal{K}$ is uniquely determined by its support function

$$h(K, u) := \max\{\langle x, u \rangle : x \in K\}, \quad u \in S^{d-1}.$$

The Steiner point of $K \in \mathcal{K}$ is given by

$$s(K) := \kappa_d^{-1} \int_{S^{d-1}} h(K, u)\, u\, \sigma(du).$$

The centred support function of $K \in \mathcal{K}$ is

$$h^*(K, u) := h(K - s(K), u), \quad u \in S^{d-1},$$

and is an additive functional on $\mathcal{K}$.

For $0 \leq k \leq d$ the set of all k-faces of a polytope $P \subset \mathbb{R}^d$ is denoted by $\mathcal{F}_k(P)$. The orthogonal group is denoted by O_d. The group of proper rotations is denoted by SO_d and is equipped with its standard topology. We denote the identity in SO_d by id. The unique normalized Haar measure on SO_d is denoted by ν. For the unit sphere, the k-fold product of the unit sphere and the rotation group the L^p-spaces will be equipped with the corresponding Haar probability measure:

$$L^p(S^{d-1}) := L^p\left(S^{d-1}, \omega_d^{-1}\sigma\right), \; L^p\left((S^{d-1})^k\right) := L^p\left((S^{d-1})^k, \omega_d^{-k}\sigma^k\right)$$

and

$$L^p(SO_d) := L^p(SO_d, \nu).$$

We define by $c : \mathcal{C} \to \mathbb{R}^d$ the mapping that associates with each $C \in \mathcal{C}$ the center of the (uniquely determined) smallest ball containing C. The mapping c is continuous with respect to the Hausdorff metric; see [SW08, Lemma 4.1.1]. Furthermore, we define the grain space $\mathcal{C}_0 := \{C \in \mathcal{C} : c(C) = 0\}$ and correspondingly $\mathcal{K}_0 := \mathcal{C}_0 \cap \mathcal{K}$ and $\mathcal{R}_0 := \mathcal{C}_0 \cap \mathcal{R}$.

For $R \in \mathcal{R}$, we define $N(R) := \min\{m \in \mathbb{N} : R = \bigcup_{i=1}^m K_i \text{ with } K_i \in \mathcal{K}\}$ and $N(\emptyset) := 0$. The function $N : \mathcal{R} \cup \{\emptyset\} :\to \mathbb{N}_0$ is measurable, compare [SW08, Lemma 4.3.1].

Let $L_1, \dots, L_k \subset \mathbb{R}^d$ be linear subspaces with $\dim L_1 + \dots + \dim L_k =: m \leq d$. Then we choose an orthonormal basis in each subspace L_j and define $\det(L_1, \dots, L_k)$ as the m-dimensional volume of the parallelepiped which is spanned by the union of these orthonormal bases. On the other hand, if $\dim L_1 + \dots + \dim L_k \geq (k-1)d$, we define

$$[L_1, \dots, L_k] := \det(L_1^\perp, \dots, L_k^\perp).$$

Moreover, if $A_1, \dots, A_k$ are non-empty convex sets with $\dim A_1 + \dots + \dim A_k \geq (k-1)d$ and $L(A_i)$ denotes the linear subspace which is parallel to the affine hull of A_i, then we define

$$[A_1, \dots, A_k] := [L(A_1), \dots, L(A_k)].$$

2.3 Foundations from stochastic geometry

In this section we introduce random closed sets, point processes and particle processes. For more information on these fundamental notions of stochastic geometry we refer to the survey article [SW10] and to the monographs [SW08] and [CSKM13]. In the subsequent section the notions are combined and used to introduce the main object of research in this thesis: the Boolean model. It is the important random closed set which is obtained as the union set of a Poisson particle process.

We denote by $\mathcal{F}$ the class of closed subsets of $\mathbb{R}^d$ and for $A \subset \mathbb{R}^d$ by $\mathcal{F}_A$ the class of closed sets which hit A and by $\mathcal{F}^A$ the class of closed sets which miss A. We equip $\mathcal{F}$ with the hit-or-miss topology, which is generated by the system

$$\{\mathcal{F}^C : C \in \mathcal{C}\} \cup \{\mathcal{F}_G : G \subset \mathbb{R}^d \text{ open}\}.$$

An $\mathcal{F}$-valued random variable is called a random closed set. On the subclass $\mathcal{C}$ of compact sets the Borel σ-algebra induced by the Hausdorff metric coincides with the one induced from the hit-or-miss topology. A random closed set Z with a distribution $\mathbb{P}_Z$ which is invariant under translations of Z is called stationary. If $\mathbb{P}_Z$ is invariant under rotations of Z, the random closed set Z is called isotropic.

Let E be a locally compact space with countable base. A counting measure η on E is a sum of Dirac measures

$$\eta := \sum_{i=1}^{k} \delta_{x_i}, \quad k \in \mathbb{N}_0 \cup \{\infty\}, x_i \in E, 1 \leq i \leq k.$$

If the points $x_i, i \in \mathbb{N}$, are distinct we call η simple. A simple counting measure η can be identified with its support $\{x_1, x_2, \ldots\}$. For $x \in E$ it makes sense to write $x \in \eta$ if $\eta(\{x\}) = 1$. We call η locally finite if $\eta(C) < \infty$ for compact sets $C \subset E$. We denote by $N_s(E)$ the collection of locally finite simple counting measures. For $A \in \mathcal{B}(E)$ let $N_{s,A}$ be the collection of locally finite simple counting measures without points in A. Then, we equip $N_s(E)$ with the σ-algebra $\mathcal{N}_s(E)$ which is generated by the system

$$\{N_{s,G} : G \subset E \text{ open and relatively compact}\}.$$

A point process X on E is a random variable with values in the space $N_s(E)$ of simple counting measures on E. A point process on $\mathbb{R}^d$ can be identified with its support and can thus be interpreted as a locally finite random closed set. For $E = \mathbb{R}^d$ or $E = \mathcal{F} \setminus \{\emptyset\}$ we define rotations and translations of a point process X by pointwise application. X is called isotropic if its distribution $\mathbb{P}_X$ is invariant under rotations of X and stationary if $\mathbb{P}_X$ is invariant under translations of X. The intensity measure of a point process X is defined by

$$\Theta(A) = \mathbb{E}X(A), \quad A \in \mathcal{B}(E).$$

It maps a Borel set A to the mean number of points of X which lie in A. For a stationary point process X in $\mathbb{R}^d$ with locally finite intensity measure Θ it holds $\Theta = \gamma\lambda$ with a constant $\gamma \geq 0$, which is called the intensity. The intensity γ is the mean number of points of X per unit volume.

A point process in $\mathcal{F} \setminus \{\emptyset\}$ whose intensity measure is concentrated on $\mathcal{C}$ is called a particle process. In this thesis we make the general assumption that the intensity measure Θ of a particle process is locally finite, which is equivalent to the condition

$$\Theta(\mathcal{F}_C) < \infty, \quad C \in \mathcal{C}.$$

Then, for a stationary particle process X with intensity measure $\Theta \neq 0$ there exist a unique number $\gamma > 0$ and a unique probability measure $\mathbb{Q}$ on $\mathcal{C}_0$ such that

$$\mathbb{E}X(A) = \gamma \int_{\mathcal{C}_0} \int_{\mathbb{R}^d} \mathbf{1}\{C + x \in A\} dx\, \mathbb{Q}(dC), \quad A \in \mathcal{B}(\mathcal{C}). \tag{2.3}$$

We call γ the intensity and $\mathbb{Q}$ the grain distribution of X. The right-hand side of (2.3) defines a locally finite measure if and only if the grain distribution fulfils the integrability condition

$$\int_{\mathcal{C}_0} \lambda(C + B^d) \mathbb{Q}(dC) < \infty. \tag{2.4}$$

Therefore, we assume that the condition (2.4) is fulfilled for all grain distributions occurring in the following. The random compact set Z_0 with distribution $\mathbb{Q}$ is called the typical grain of X. If X is a particle process with convex particles, (2.4) is by the Steiner formula (2.1) equivalent to the $\mathbb{Q}$-integrability of the intrinsic volumes $V_0, \ldots, V_d$. For a measurable, translation invariant functional φ on $\mathcal{C}$ which is non-negative or $\mathbb{Q}$-integrable we define the φ-density of X as

$$\overline{\varphi}(X) := \gamma \mathbb{E} \varphi(Z_0).$$

The scaling of the expectation by the intensity is motivated by the relations

$$\overline{\varphi}(X) = \frac{1}{\lambda(B)} \mathbb{E} \sum_{C \in X, c(C) \in B} \varphi(C) = \lim_{r \to \infty} \frac{1}{\lambda(rW)} \mathbb{E} \sum_{C \in X, C \subset rW} \varphi(C), \tag{2.5}$$

where $B \subset \mathbb{R}^d$ is a Borel set with $0 < \lambda(B) < \infty$ and $W \in \mathcal{K}$ is a convex body with $\lambda(W) > 0$.

For a particle process X satisfying (2.4) the union set

$$Z_X := \bigcup_{C \in X} C$$

is a random closed set by [SW10, Theorem 1.22].

A Poisson process in E is a point process X with the property that $X(A)$ is Poisson distributed with parameter $\Theta(A) = \mathbb{E}X(A)$ for all Borel sets $A \subset E$ with $\Theta(A) < \infty$. If Θ is a locally finite measure on E without atoms there exists a in distribution unique Poisson process with intensity measure Θ (see [SW08, Theorem 3.2.1]). A Poisson process X has independent increments, that is for pairwise disjoint Borel sets $A_1, \ldots, A_k \subset E$ with $\Theta(A_i) < \infty, 1 \leq i \leq k$ the random variables $X(A_1), \ldots, X(A_k)$ are stochastically independent.

Let X be a Poisson particle process with intensity γ and grain distribution $\mathbb{Q}$ satisfying the integrability condition (2.4). Then, the union set

$$Z := \bigcup_{K \in X} K$$

is called a Boolean model. In this thesis we assume that X is a stationary Poisson process of convex particles. A functional φ on $\mathcal{R}$ which is translation invariant, additive, measurable and bounded on $\{K \in \mathcal{K} : K \subset C^d\}$ is called geometric. Assume we observe the stationary Boolean model Z in an observation window $W \in \mathcal{K}$ with $V_d(W) > 0$. Then, the mean value of φ applied to the intersection of the Boolean model with the observation window is given by

$$\mathbb{E}\varphi(Z \cap W) = \sum_{k=1}^{\infty} \frac{(-1)^{k-1}}{k!} \int_{(\mathcal{K}_0)^k} \int_{(\mathbb{R}^d)^k} \varphi(W \cap (K_1 + x_1) \cap \ldots \\ \ldots \cap (K_k + x_k)) d(x_1, \ldots, x_k) \mathbb{Q}^k(d(K_1, \ldots, K_k)), \tag{2.6}$$

see [SW08, Theorem 9.1.2]. It can be shown that the series in (2.6) converges absolutely. Furthermore, the limit

$$\overline{\varphi}(Z) := \lim_{r \to \infty} \frac{\mathbb{E}\varphi(Z \cap rW)}{V_d(rW)} \tag{2.7}$$

is independent of W and defines the φ-density of the Boolean model Z.

Formula (2.6) and a calculation using essentially Fubini's theorem imply the density formulas for V_d and V_{d-1} stated in (1.2).

We put our focus on non-isotropic Boolean models where densities of so-called mixed geometric functionals depending on several convex bodies play an important role. For a measurable, translation invariant functional $\varphi : \mathcal{K}^k \to \mathbb{R}$ which is $\mathbb{Q}^k$-integrable or nonnegative the φ-density of X is defined by

$$\overline{\varphi}(\underbrace{X, \ldots, X}_{k \text{ times}}) := \gamma^k \int_{\mathcal{K}_0} \varphi(K_1, \ldots, K_k) \mathbb{Q}^k(d(K_1, \ldots, K_k))$$

For example, in the density formulas for the intrinsic volumes $V_{d-2}, \ldots, V_0$, the densities of X for special mixed functionals occur. In this thesis we consider densities (or specific values) for various geometric functionals and even for measures which are, if evaluated at a fixed Borel set, geometric functionals.

2.4 Spherical harmonics and two auxiliary lemmas

In this section we briefly recall some results on spherical harmonics. More detailed information on this topic can be found for example in [DX13, Chapter 1].

Denote by ∂_i, $1 \leq i \leq d$ the ith partial derivative and by $\Delta := \partial_1^2 + \ldots + \partial_d^2$ the Laplace operator on $\mathbb{R}^d$. A spherical polynomial is the restriction of a polynomial on $\mathbb{R}^d$ to the unit sphere. For $n \in \mathbb{N}_0$ define by $\Pi_n(S^{d-1})$ the space of spherical polynomials of degree at most n. A spherical harmonic is a homogeneous spherical polynomial q with $\Delta q = 0$. Denote by $\mathcal{S}_n$ the space of spherical harmonics of polynomial degree n. By [DX13, Corollary 1.14] the space $\mathcal{S}_n$ has the dimension

$$D(d,n) = \binom{n+d-1}{n} - \binom{n+d-3}{n-2},$$

where the binomial coefficient is defined as 0 if the lower entry is negative. The spaces $\mathcal{S}_j$ and $\mathcal{S}_k$ for $j \neq k$ are orthogonal subspaces of $L^2(S^{d-1})$. Furthermore, we have

$$\Pi_n(S^{d-1}) = \mathcal{S}_0 \oplus \ldots \oplus \mathcal{S}_n. \tag{2.8}$$

For every $l \in \mathbb{N}_0$ we can choose an orthonormal basis $B_l := \{Y_{l,1}, \ldots, Y_{l,D(d,l)}\}$ of $\mathcal{S}_l$. Furthermore, $B := \bigcup_{l=0}^{\infty} B_l$ is an orthonormal basis of $L^2(S^{d-1})$. This implies the identity

$$(f,g) = \sum_{Y \in B} (f,Y)(Y,g), \quad f,g \in L^2(S^{d-1}). \tag{2.9}$$

In two and three dimensions an orthonormal basis of the L^2-integrable functions on the unit sphere can be stated explicitly. For $d = 2$ we have

$$D(2,l) = \begin{cases} 1, & l = 0, \\ 2, & l > 0, \end{cases}$$

and $B_0 = \{Y_{0,1}\}$ and $B_l = \{Y_{l,1}, Y_{l,2}\}$ with

$$\begin{aligned} Y_{0,1} &\equiv 1, \\ Y_{l,1}(u(\theta)) &= \cos(l\theta), \\ Y_{l,2}(u(\theta)) &= \sin(l\theta), \end{aligned}$$

where $l \in \mathbb{N}$, $u(\theta) := (\cos\theta, \sin\theta)^\top$ and $\theta \in [0, 2\pi)$.

For $d = 3$ and $l \in \mathbb{N}_0$ we have $D(3,l) = 2l + 1$. An orthonormal basis can be stated explicitly using the Gegenbauer polynomials

$$C_l^\lambda, \quad \lambda > -1/2, l \in \mathbb{N}_0,$$

see [DX13, Appendix B.2] for their properties. Namely, an orthonormal basis is given by $B_l = \{Y_{l,2k} : 1 \leq k \leq l\} \cup \{Y_{l,2k+1} : 0 \leq k \leq l\}$ with

$$\begin{aligned} Y_{l,2k}(u(\theta,\phi)) &= (\sin\theta)^k C_{l-k}^{k+\frac{1}{2}}(\cos\theta)\sin(k\phi), \quad 1 \leq k \leq l, \\ Y_{l,2k+1}(u(\theta,\phi)) &= (\sin\theta)^k C_{l-k}^{k+\frac{1}{2}}(\cos\theta)\cos(k\phi), \quad 0 \leq k \leq l, \end{aligned}$$

where $l \in \mathbb{N}_0, u(\theta,\phi) := (\sin\theta\sin\phi, \sin\theta\cos\phi, \cos\theta)^\top$ and $0 \leq \theta \leq \pi$, $0 \leq \phi \leq 2\pi$. The tensor product of $f_1, \ldots, f_k : S^{d-1} \to \mathbb{R}$ is the function $f_1 \otimes \ldots \otimes f_k : (S^{d-1})^k \to \mathbb{R}$ which maps $(x_1, \ldots, x_k) \in (S^{d-1})^k$ to $f_1(x_1) \cdots f_k(x_k)$. On the product space $(S^{d-1})^k$ the tensor product is a useful tool to obtain an orthonormal basis and a representation of the scalar product.

Lemma 2.4.1. *For every $k \in \mathbb{N}$ the set*

$$\{Y_1 \otimes \ldots \otimes Y_k : Y_1, \ldots, Y_k \in B\} \tag{2.10}$$

is an orthonormal basis of $L^2((S^{d-1})^k)$ and

$$(f,g) = \sum_{Y_1,\ldots,Y_k \in B} (f, Y_1 \otimes \ldots \otimes Y_k)(Y_1 \otimes \ldots \otimes Y_k, g), \quad f,g \in L^2((S^{d-1})^k). \tag{2.11}$$

Proof. It is a well-known Hilbert space property that equation (2.11) is fulfilled if (2.10) forms an orthonormal basis of $L^2((S^{d-1})^k)$.

By [Mac09, Theorem 1.33] this is the case if for all $f \in L^2((S^{d-1})^k)$ the property

$$(f, Y_1 \otimes \ldots \otimes Y_k) = 0, \qquad \text{for all } Y_1, \ldots, Y_k \in B$$

implies

$$f(x_1, \ldots, x_k) = 0 \text{ for } \sigma^k\text{-almost all } (x_1, \ldots, x_k) \in (S^{d-1})^k. \tag{2.12}$$

We proceed by induction. For $k = 1$ the assertion (2.12) holds obviously since B is an orthonormal basis of $L^2(S^{d-1})$. Now we assume for some $k \in \mathbb{N}$ that (2.12) is fulfilled for all $f \in L^2((S^{d-1})^k)$. Let $f \in L^2((S^{d-1})^{k+1})$. We define

$$f_z(x_1, \ldots, x_k) := f(x_1, \ldots, x_k, z), \quad x_1, \ldots, x_k, z \in S^{d-1}.$$

Then, $f_z \in L^2\left(\left(S^{d-1}\right)^k\right)$ for σ-almost all $z \in S^{d-1}$. Assume

$$(f, Y_1 \otimes \ldots \otimes Y_{k+1}) = \omega_d^{-1} \int\limits_{S^{d-1}} (f_z, Y_1 \otimes \ldots \otimes Y_k) Y_{k+1}(z) \sigma(dz) = 0$$

for all $Y_1, \ldots, Y_{k+1} \in B$.

Then, it follows for all $Y_1, \ldots, Y_k \in B$ that

$$(f_z, Y_1 \otimes \ldots \otimes Y_k) = 0 \text{ for } \sigma\text{-almost all } z \in S^{d-1}.$$

This yields $f(x_1, \ldots, x_{k+1}) = 0$ for σ^{k+1}-almost all $(x_1, \ldots, x_{k+1}) \in (S^{d-1})^{k+1}$ and thus the assertion.

□

For $f \in L^2(S^{d-1})$ denote by $\pi_n f$ the orthogonal projection on $\Pi_n(S^{d-1})$. For $k \in \mathbb{N}$ and $f \in L^2((S^{d-1})^k)$ we denote by $\pi_n f$ the orthogonal projection on the space

$$\{f_1 \otimes \ldots \otimes f_k : f_1, \ldots, f_k \in \Pi_n(S^{d-1})\}.$$

The following lemma provides an upper bound for the L^2-error which arises if tensor products of functions on the unit sphere are approximated by spherical polynomials.

Lemma 2.4.2. *Let $k \in \mathbb{N}$ and $g_1, \dots, g_k \in L^2(S^{d-1})$. Then*

$$\|g_1 \otimes \ldots \otimes g_k - \pi_n(g_1 \otimes \ldots \otimes g_k)\|_2 \leq \sum_{i=1}^{k} \|g_i - \pi_n g_i\|_2 \prod_{\substack{j=1 \\ j \neq i}}^{k} \|g_j\|_2.$$

Proof. We have

$$\pi_n(g_1 \otimes \ldots \otimes g_k) = \pi_n(g_1) \otimes \ldots \otimes \pi_n(g_k).$$

Thus, we get

$$\begin{aligned}
&\|g_1 \otimes \ldots \otimes g_k - \pi_n(g_1 \otimes \ldots \otimes g_k)\|_2 \\
&= \|g_1 \otimes \ldots \otimes g_k - \pi_n(g_1) \otimes \ldots \otimes \pi_n(g_k)\|_2 \\
&\leq \|(g_1 - \pi_n(g_1)) \otimes g_2 \otimes \ldots \otimes g_k\|_2 + \|\pi_n(g_1) \otimes (g_2 - \pi_n(g_2)) \otimes g_3 \otimes \ldots \otimes g_k)\|_2 + \ldots \\
&\quad \ldots + \|\pi_n(g_1) \otimes \ldots \otimes \pi_n(g_{k-1}) \otimes (g_k - \pi_n(g_k))\|_2 \\
&\leq \sum_{i=1}^{k} \|g_i - \pi_n(g_i)\|_2 \prod_{\substack{j=1 \\ j \neq i}}^{k} \|g_j\|_2,
\end{aligned}$$

which yields the assertion. □

By [DX13, Lemma 1.2.4] the projection operator π_n has an integral representation using a reproducing kernel given by

$$Z_n(x, y) = \sum_{Y \in B_n} Y(x)Y(y), \quad x, y \in S^{d-1}.$$

Namely, for $f \in L^2(S^{d-1})$ it holds

$$[\pi_n f](x) = \omega_d^{-1} \int_{S^{d-1}} f(y) Z_n(x, y) \sigma(dy), \quad x \in S^{d-1}. \tag{2.13}$$

An easy calculation shows that

$$(Z_n(x, \cdot), Z_n(y, \cdot)) = Z_n(x, y), \quad x, y \in S^{d-1}. \tag{2.14}$$

By [DX13, Lemma 1.2.7] the reproducing kernel has for $d \geq 3$ the properties

$$|Z_n(x, y)| \leq D(d, n) \text{ and } Z_n(x, x) = D(d, n), \quad x, y \in S^{d-1}. \tag{2.15}$$

Lemma 2.4.3. *Let $n \in \mathbb{N}_0$ and $Y \in \mathcal{S}_n$. Then*

$$\|Y\|_\infty \leq \|Y\|_2 \sqrt{D(d, n)}.$$

Proof. For all $x \in S^{d-1}$ we obtain by (2.13) and from the Cauchy-Schwarz inequality that

$$|Y(x)| = |[\pi_n Y](x)| = |(Y, Z_n(x,\cdot))| \leq \|Y\|_2 \|Z_n(x,\cdot)\|_2.$$

By (2.14) and (2.15) we obtain

$$\|Z_n(x,\cdot)\|_2 = \sqrt{(Z_n(x,\cdot), Z_n(x,\cdot))} = \sqrt{Z_n(x,x)} = \sqrt{D(d,n)},$$

which implies the assertion. □

2.5 A decomposition of the Haar measure on the rotation group

For linearly independent vectors $v_1, \ldots, v_d \in S^{d-1}$ we denote by $GS(v_1, \ldots, v_d)$ a proper rotation obtained by applying the Gram-Schmidt orthonormalization. More precisely, let $GS(v_1, \ldots, v_d)$ be the unique matrix in SO_d with columns $z_1, \ldots, z_d$ fulfilling the recursion

$$y_k := v_k - \sum_{i=1}^{k-1} \langle v_k, y_i \rangle \frac{y_i}{\|y_i\|^2}, \quad 1 \leq k \leq d,$$

$$z_k := y_k / \|y_k\|, \quad 1 \leq k \leq d-1 \text{ and } z_d \in \left\{ \pm \frac{y_d}{\|y_d\|} \right\},$$

where the sign of z_d is chosen such that the determinant of $GS(v_1, \ldots, v_d)$ is equal to 1. Note that

$$GS(\vartheta v_1, \ldots, \vartheta v_d) = \vartheta\, GS(v_1, \ldots, v_d), \quad \vartheta \in SO_d, v_1, \ldots, v_d \in S^{d-1}. \tag{2.16}$$

Fix some $\vartheta_0 \in SO_d$. We define a mapping $\Psi : (S^{d-1})^d \to SO_d$ by

$$\Psi(v_1, \ldots, v_d) = \begin{cases} GS(v_1, \ldots, v_d), & \text{for linearly independent } v_1, \ldots, v_d, \\ \vartheta_0, & \text{otherwise.} \end{cases}$$

For $v \in S^{d-1}$ and $k := \min_{1 \leq i \leq d} \{ i : \langle v, e_i \rangle \neq 0 \}$ let

$$\vartheta_v := \Psi(v, e_1, \ldots, e_{k-1}, e_{k+1}, \ldots, e_d). \tag{2.17}$$

Note that $\vartheta_v e_1 = v$ for all $v \in S^{d-1}$. Define by $SO_d(v) := \{ \vartheta \in SO_d : \vartheta v = v \}$ the set of all rotations leaving v fixed. Since $SO_d(v)$ is a compact topological group with countable basis, there exists a unique Haar probability measure ν_v on $SO_d(v)$. For a proof of this general result compare e.g. [SW08, Theorem 13.1.3].

By (2.16) we get for $\vartheta \in SO_d(e_1)$ and σ^{d-1}-almost all $(v_2, \dots, v_d) \in \left(S^{d-1}\right)^{d-1}$ that

$$\vartheta\Psi(e_1, v_2, \dots, v_d) = \Psi(\vartheta e_1, \vartheta v_2, \dots, \vartheta v_d) = \Psi(e_1, \vartheta v_2, \dots, \vartheta v_d).$$

This and the rotation invariance of σ imply the explicit representation of the Haar probability measure on $SO_d(e_1)$ as

$$\nu_{e_1}(B) = \frac{1}{\omega_d^{d-1}} \int\limits_{(S^{d-1})^{d-1}} \mathbf{1}\{\Psi(e_1, v_2, \dots, v_d) \in B\}\sigma^{d-1}(d(v_2, \dots, v_d)), \tag{2.18}$$

for Borel sets $B \subset SO_d(e_1)$. Now, we obtain a representation of the Haar measure on the rotation group as an image measure with respect to the product of the Haar measures on the unit sphere and the group of rotations leaving e_1 fixed.

Lemma 2.5.1. *We have*

$$\nu(A) = \frac{1}{\omega_d} \int\limits_{S^{d-1}} \int\limits_{SO_d(e_1)} \mathbf{1}\{\vartheta_v \vartheta \in A\}\nu_{e_1}(d\vartheta)\sigma(dv), \quad A \in \mathcal{B}(SO_d).$$

Proof. From the proof of [SW08, Theorem 13.2.9], by (2.16), the rotation invariance of σ and (2.18) we obtain that

$$\begin{aligned}
\nu(A) &= \frac{1}{\omega_d^d} \int\limits_{(S^{d-1})^d} \mathbf{1}\{\Psi(v, v_2, \dots, v_d) \in A\}\sigma^d(d(v, v_2, \dots, v_d)) \\
&= \frac{1}{\omega_d^d} \int\limits_{(S^{d-1})^d} \mathbf{1}\{\vartheta_v \Psi(e_1, \vartheta_v^{-1} v_2, \dots, \vartheta_v^{-1} v_d) \in A\}\sigma^d(d(v, v_2, \dots, v_d)) \\
&= \frac{1}{\omega_d} \int\limits_{S^{d-1}} \int\limits_{SO(e_1)} \mathbf{1}\{\vartheta_v \vartheta \in A\}\nu_{e_1}(d\vartheta)\sigma(dv),
\end{aligned}$$

which completes the proof. □

Remark 2.5.2. In Lemma 2.5.1 we obtain a disintegration of the Haar measure on the group SO_d with respect to the Haar measure on the unit sphere S^{d-1} by an explicit construction. This disintegration can be considered as a special case of a much more general result. Let G be a locally compact second countable Hausdorff topological group which acts properly on a topological space S. Then, Gentner and Last showed in [GL11, (11)] that the Haar measure on G has an invariant disintegration along each orbit in S. The special case where G acts transitively on S, which is dealt with in [GL11, Example 1], applies to the situation of Lemma 2.5.1.

3 Translative integral geometric results for support and area measures

3.1 Support and area measures

For a convex body $K \in \mathcal{K}$ and $x \in \mathbb{R}^d$, let $p(K,x)$ denote the metric projection of x to K. For $x \in K$ we define the normal cone of K at x by $N(K,x) := \{u \in \mathbb{R}^d : p(K, x+u) = x\}$, and for nonempty, convex $F \subset K$ let $N(K,F) := N(K,x)$, where $x \in \operatorname{relint} F$. For $x \notin K$ put $u(K,x) := (x - p(K,x))/\|x - p(K,x)\|$. For $K \in \mathcal{K}$ denote by

$$\operatorname{Nor}(K) := \{(x,u) \in \partial K \times S^{d-1} : u \in N(K,x)\}$$

the normal bundle of K. Then, the support measures (generalized curvature measures) $\Xi_0(K,\cdot), \dots, \Xi_{d-1}(K,\cdot)$ of a convex body $K \in \mathcal{K}$ are defined by a local Steiner formula. Namely, for any $\epsilon > 0$ and Borel set $\eta \subset \Sigma := \mathbb{R}^d \times S^{d-1}$, the d-dimensional Hausdorff measure (volume) of the local parallel set

$$M_\epsilon(K,\eta) := \{x \in (K + \epsilon B^d) \setminus K : (p(K,x), u(K,x)) \in \eta\}$$

is a polynomial in ϵ, that is,

$$\mathcal{H}^d(M_\epsilon(K,\eta)) = \sum_{j=0}^{d-1} \epsilon^{d-j} \kappa_{d-j} \Xi_j(K,\eta);$$

see [Sch13a, SW08] for further information. The support measures are related to the intrinsic volumes by $V_j(K) = \Xi_j(K,\Sigma)$, for $j = 0, \dots, d-1$. The marginal measures of the jth support measure are the measure

$$\Psi_j(K,B) := \Xi_j(K, \mathbb{R}^d \times B), \quad B \in \mathcal{B}(S^{d-1}),$$

which is called the jth area measure of K and

$$\Phi_j(K,A) := \Xi_j(K, A \times S^{d-1}), \quad A \in \mathcal{B}(\mathbb{R}^d),$$

which is called the jth curvature measure of K. Furthermore, it is convenient to extend the definition of the support measures to $K = \emptyset$ by

$$\Xi_j(\emptyset, \cdot) = 0.$$

This convention is used for all convex body-valued functionals derived from the support measures which will be introduced in this thesis. Thus, we have

$$\Psi_j(\emptyset, \cdot) = 0, \quad \Phi_j(\emptyset, \cdot) = 0 \quad \text{and} \quad V_j(\emptyset) = 0.$$

The support measures and the additive functionals derived from them have a unique additive extension to the convex ring $\mathcal{R}$. The following theorem collects some important properties of the support measures.

Theorem 3.1.1. *Let $j \in \{0, \dots, d-1\}$, $K \in \mathcal{R}$ and let $A \subset \mathbb{R}^d$ and $C \subset S^{d-1}$ be Borel sets.*

(i) *Positive Homogineity: For $\lambda > 0$ we have*

$$\Xi_j(\lambda K, \lambda A \times C) = \lambda^j \, \Xi_j(K, A).$$

(ii) *Additivity: The map $K \mapsto \Xi_j(K, \cdot)$ is additive on $\mathcal{R}$.*

(iii) *Weak continuity: The map $K \mapsto \Xi_j(K, \cdot)$ is weakly continuous on $\mathcal{K}$.*

(v) *Translation covariance:*

$$\Xi_j(K + x, (A + x) \times C) = \Xi_j(K, A \times C), \quad x \in \mathbb{R}^d, K \in \mathcal{K}.$$

(vi) *Rotation covariance:*

$$\Xi_j(\vartheta K, (\vartheta A) \times (\vartheta C)) = \Xi_j(K, A \times C), \quad \vartheta \in SO_d, K \in \mathcal{K}.$$

For special convex bodies the support measures have an explicit representation. Namely, for a polytope P the jth support measure can be represented as a sum over the j-faces of P by [Sch13a, (4.21)]. On the other hand, by [Sch13a, (4.25)] for a smooth convex body $K \in C^2_+$ the jth curvature measure has a representation as an integral over the elementary symmetric function H_{d-1-j} of the principal curvatures of ∂K and by [Sch13a, (4.26)] the jth area measure has a representation as integral of the jth normalized elementary symmetric function s_j of the principal radii of curvature of ∂K.

Lemma 3.1.2. *Let $j \in \{0, \dots, d-1\}$ and $A \subset \mathbb{R}^d$ and $C \subset S^{d-1}$ be Borel sets.*

(i) *If P is a polytope, then*

$$\Xi_j(P, A \times C) = \frac{1}{\omega_{d-j}} \sum_{F \in \mathcal{F}_j(P)} \mathcal{H}^{d-1-j}(N(P, F) \cap C) \, \mathcal{H}^j(F \cap A).$$

(ii) *If $K \in C^2_+$ is a smooth convex body, then*

$$\Psi_j(K,C) = \frac{\binom{d-1}{j}}{\omega_{d-j}} \int\limits_{S^{d-1}} \mathbf{1}\{u \in C\} s_j(u) \sigma(du)$$

and

$$\Phi_j(K,A) = \frac{\binom{d-1}{j}}{\omega_{d-j}} \int\limits_{\partial K} \mathbf{1}\{x \in A\} H_{d-1-j}(x) \mathcal{H}^{d-1}(dx).$$

3.2 Translative integral formulas

In the following, a crucial role will be played by an iterated translative integral formula for support measures which is shown in [Hug99, Theorem 3.14].

3.2.1 The translative integral formula

For $j \in \{0,\dots,d\}$ and $k \in \mathbb{N}$, we define a collection of multi-indices by

$$\operatorname{mix}(j,k) := \{(m_1,\dots,m_k) \in \{j,\dots,d\}^k : m_1 + \dots + m_k = (k-1)d + j\}.$$

For $m \in \operatorname{mix}(j,k)$, we write $\operatorname{type}(m) := j$ and $|m| := k$. For $j \le j' \le d$, $k' \in \mathbb{N}$, $m := (m_1,\dots,m_k) \in \operatorname{mix}(d-j'+j,k)$ and $m' := (m'_1,\dots,m'_{k'}) \in \operatorname{mix}(j',k')$, we write

$$(m,m') := (m_1,\dots,m_k,m'_1,\dots,m'_{k'})$$

for the concatenation of m and m'. If the concatenation (m,m') occurs as index we usually omit the brackets.

For convex bodies $K_1,\dots,K_k \in \mathcal{K}$, $j \in \{0,\dots,d-1\}$, and $k \in \mathbb{N}$ the following iterated translative integral formula holds.

Theorem 3.2.1. *(Hug 1999)*
There exists a unique collection of Borel measures

$$\Xi^{(j)}_m(K_1,\dots,K_k;\cdot), \quad m := (m_1,\dots,m_k) \in \operatorname{mix}(j,k)$$

on $(\mathbb{R}^d)^k \times S^{d-1}$ such that $\Xi^{(j)}_{m_1,\dots,m_k}(K_1,\dots,K_k;A_1\times\dots\times A_k\times C)$ is positively homogeneous of degree m_i with respect to (K_i,A_i) and

$$\int\limits_{(\mathbb{R}^d)^{k-1}} \Xi_j(K_1\cap(K_2+x_2)\cap\dots\cap(K_k+x_k),(A_1\cap(A_2+x_2)\cap\dots$$

$$\ldots \cap (A_k + x_k)) \times C)d(x_2, \ldots, x_k)$$
$$= \sum_{m \in \text{mix}(j,k)} \Xi_m^{(j)}(K_1, \ldots, K_k; A_1 \times \ldots \times A_k \times C)$$

for all Borel sets $A_1, \ldots, A_k \subset \mathbb{R}^d$ *and* $C \subset S^{d-1}$.

Proof. For the proof we refer to [Hug99, Theorem 3.14], which states a corresponding formula in the more general framework of relative support measures. The idea is to prove the statement for polytopes by making use of their facial structure and to deduce the result for arbitrary convex bodies by approximation. □

We remark that a corresponding formula in the setting of sets with positive reach is shown in [Rat97]. The measures $\Xi_m^{(j)}(K_1, \ldots, K_k; \cdot)$, $\quad m \in \text{mix}(j,k)$, are called *mixed support measures*. Since it holds that $\Xi_j^{(j)} = \Xi_j$, the exponent (j) is redundant in the notation $\Xi_m^{(j)}$ if it is clear that $m \in \text{mix}(j,k)$. To keep the notation as simple as possible, we omit it in the following.

The special case of Theorem 3.2.1 with $A_1 = \ldots = A_k = \mathbb{R}^d$ leads to a translative integral formula for the area measure Ψ_j with the mixed area measures of translative integral geometry

$$\Psi_m(K_1, \ldots, K_k; C) := \Xi_m(K_1, \ldots, K_k; (\mathbb{R}^d)^k \times C), \quad C \in \mathcal{B}(S^{d-1}), m \in \text{mix}(j,k)$$

on the right-hand side. The mixed area measures of translative integral geometry should be distinguished from the measures on the unit sphere which are introduced as coefficients in the polynomial expansion of the $(d-1)$th area measure of a linear combination of convex bodies and which are also called mixed area measures, see [Sch13a, Chapter 5]. If we choose $A_1 = \ldots = A_k = \mathbb{R}^d$ and $C = S^{d-1}$ in Theorem 3.2.1 we obtain a translative integral formula for the intrinsic volume V_j with the mixed functionals of translative integral geometry

$$V_m(K_1, \ldots, K_k) := \Xi_m(K_1, \ldots, K_k; (\mathbb{R}^d)^k \times S^{d-1}), \quad m \in \text{mix}(j,k)$$

on the right-hand side.

3.2.2 Properties of mixed support measures

We collect the important properties of the mixed support measures in a theorem.

Theorem 3.2.2. *Let* $m \in \text{mix}(j,k)$ *and* $A_1, \ldots, A_k \subset \mathbb{R}^d$, $C \subset S^{d-1}$, $D' \subset (\mathbb{R}^d)^{k-1} \times S^{d-1}$ *and* $D \subset (\mathbb{R}^d)^k \times S^{d-1}$ *be Borel sets.*

(i) *Symmetry:*

$\Xi_{m_1,\ldots,m_k}(K_1,\ldots,K_k;A_1\times\ldots\times A_k\times C)$ *is symmetric with respect to permutations of* $\{1,\ldots,k\}$.

(ii) *Decomposability:*

$$\Xi_{d,m_2,\ldots,m_k}(K_1,\ldots,K_k;A_1\times D')=\mathcal{H}^d(K_1\cap A_1)\Xi_{m_2,\ldots,m_k}(K_2,\ldots,K_k;D').$$

(iii) *Nonnegativity and support:* $\Xi_{m_1,\ldots,m_k}(K_1,\ldots,K_k;\cdot)$ *is a finite nonnegative Borel measure on* $(\mathbb{R}^d)^k\times S^{d-1}$*, which is supported by* $S_1\times\ldots\times S_k\times S^{d-1}$*, where* $S_i=K_i$ *if* $m_i=d$*, and* $S_i=\partial K_i$ *otherwise.*

(iv) *Positive homogeneity: For* $\lambda_1,\ldots,\lambda_k>0$ *we have*

$$\begin{aligned}&\Xi_{m_1,\ldots,m_k}(\lambda_1K_1,\ldots,\lambda_kK_k;\lambda_1A_1\times\ldots\times\lambda_kA_k\times C)\\&=\lambda_1^{m_1}\cdots\lambda_k^{m_k}\,\Xi_{m_1,\ldots,m_k}(K_1,\ldots,K_k;A_1\times\ldots\times A_k\times C).\end{aligned}$$

(v) *Representation in the case of polytopes:*

If $K_1,\ldots,K_k$ *are polytopes, then*

$$\begin{aligned}&\Xi_{m_1,\ldots,m_k}(K_1,\ldots,K_k;A_1\times\ldots\times A_k\times C)\\&=\sum_{F_1\in\mathcal{F}_{m_1}(K_1)}\cdots\sum_{F_k\in\mathcal{F}_{m_k}(K_k)}\frac{\mathcal{H}^{d-1-j}\left(\left(\sum_{i=1}^{k}N(K_i,F_i)\right)\cap C\right)}{\omega_{d-j}}\\&\quad\times[F_1,\ldots,F_k](\mathcal{H}^{m_1}\llcorner F_1)(A_1)\cdots(\mathcal{H}^{m_k}\llcorner F_k)(A_k).\end{aligned}$$

(vi) *Additivity and weak continuity:*

The map $(K_1,\ldots,K_k)\mapsto\Xi_m(K_1,\ldots,K_k;\cdot)$ *is additive and weakly continuous on* $\mathcal{K}^k$.

(vii) *Measurability:*

The map $(K_1,\ldots,K_k)\mapsto\Xi_m(K_1,\ldots,K_k;D)$ *defined on* $\mathcal{K}^k$ *is measurable.*

(ix) *Local determination:*

If $(K_1',\ldots,K_k')\in\mathcal{K}^k$, $\beta_1,\ldots,\beta_k\subset\mathbb{R}^d$ *are open sets and* $K_i\cap\beta_i=K_i'\cap\beta_i$*, for* $i=1,\ldots,k$*, then*

$$\Xi_m(K_1,\ldots,K_k;\cdot)=\Xi_m(K_1',\ldots,K_k';\cdot)$$

on Borel subsets of $\beta_1\times\ldots\times\beta_k\times S^{d-1}$.

(x) *Translation covariance:*

$$\Xi_m(K_1 + x_1, \dots, K_k + x_k; (A_1 + x_1) \times \dots \times (A_k + x_k) \times C)$$
$$= \Xi_m(K_1, \dots, K_k; A_1 \times \dots \times A_k \times C) \text{ for } x_1, \dots, x_k \in \mathbb{R}^d.$$

(xi) *Rotation covariance:*

$$\Xi_m(\vartheta K_1, \dots, \vartheta K_k; (\vartheta A_1) \times \dots \times (\vartheta A_k) \times (\vartheta C))$$
$$= \Xi_m(K_1, \dots, K_k; A_1 \times \dots \times A_k \times C) \text{ for } \vartheta \in SO_d.$$

Proof. The property (v) of the classic support measures can be found in [Hug99, Corollary 4.10]. Property (x) follows for polytopes $P_1, \dots, P_k$ from property (v). Using algebraic induction, the weak continuity and an approximation argument it is obtained for arbitrary convex bodies. The rotation covariance follows from the representation for polytopes and the weak continuity. All other properties are stated in [Hug99, Theorem 3.14]. □

The mixed support measures Ξ_m have a unique additive extension to the convex ring $\mathcal{R}$. This follows from their weak continuity and Groemer's extension theorem ([SW08, Theorem 14.4.2]). Some of their other properties (as the translative integral formula, the symmetry, the splitting property, the positive homogeneity and the translation covariance) extend immediately to their additive extensions by the inclusion-exclusion principle. But there are some differences. Instead of being positive measures the additive extensions are signed. Also the weak continuity is lost and only the following measurability property remains.

Lemma 3.2.3. *Let $j \in \{0, \dots, d-1\}$, $m \in \operatorname{mix}(j,k)$ and $f : (\mathbb{R}^d)^k \times S^{d-1} \to [0, \infty)$ be measurable. Then*

$$(K_1, \dots, K_k) \mapsto \int_{(\mathbb{R}^d)^k \times S^{d-1}} f(x_1, \dots, x_k, u) \Xi_m(K_1, \dots, K_k; d(x_1, \dots, x_k, u))$$

is measurable on $\mathcal{R}^k$.

Proof. The additivity of the support measures implies the additivity of the considered map on $\mathcal{R}^k$. For every real-valued continuous function f on $(\mathbb{R}^d)^k \times S^{d-1}$ with compact support the above map is continuous on $\mathcal{K}^k$ by the weak continuity of Ξ_m. [SW08, Theorem 14.4.4] implies that the map is measurable on $\mathcal{R}^k$. Thus, the assertion follows from the general topological result [SW08, Lemma 12.1.1]. □

We extend the definition of the mixed support measures to locally polyconvex sets $K_1, \dots, K_k \subset \mathbb{R}^n$. Let $B \subset \mathbb{R}^d$ be a closed cube. Then, it makes sense to define by

$$\Xi_m(K_1, \dots, K_k; \cdot) := \Xi_m(K_1 \cap B, \dots, K_k \cap B; \cdot)$$

the mixed support measure of $K_1, \dots, K_k$ as a measure on $\mathcal{B}\left((\operatorname{int} B)^k \times S^{d-1}\right)$. We allow in the following the evaluation of $\Xi_m(K_1, \dots, K_k; \cdot)$ in arbitrary bounded Borel sets and assume implicitly that B is chosen large enough.

A special collection of mixed measures

In the following we need special mixed measures on the space $\left(\mathbb{R}^d\right)^k \times S^{d-1}, k \in \mathbb{N}$, which are derived from mixed support measures where some of the convex bodies are replaced by halfspaces. They will play a role in integral representations of the mixed support measures with some of the indices equal to $d-1$. The marginal measures on $\left(\mathbb{R}^d\right)^k$ of these new mixed measures are special mixed curvature measures with some of the arguments replaced by half-spaces, which have been introduced and investigated by Goodey and Weil [GW02].

For $u \in S^{d-1}$ we define by

$$u^- := \{x \in \mathbb{R}^d : \langle x, u \rangle \leq 0\}$$

the closed halfspace bounded by $u^\perp$ with outer normal u. Let

$$f(x) := (1 - \|x\|)\mathbf{1}\{\|x\| < 1\} d\kappa_{d-1}^{-1}, \quad x \in \mathbb{R}^d.$$

Using polar coordinates we can show that

$$\int_{u^\perp} f(x) \mathcal{H}^{d-1}(dx) = 1, \quad u \in S^{d-1}. \tag{3.1}$$

Then, we define special mixed measures in the following way.

Definition 3.2.4. Let $k, k' \in \mathbb{N}$, $1 \leq k \leq d-j$ and $m = (m_1, \dots, m_{k'}) \in \operatorname{mix}(k+j, k')$, $u_1, \dots, u_k \in S^{d-1}$ and $K_1, \dots, K_{k'} \in \mathcal{R}$. Then,

$$\begin{aligned} &\Xi_m(u_1, \dots, u_k, K_1, \dots, K_{k'}; \cdot) \\ &:= 2^k \int_{(\mathbb{R}^d)^{k+k'} \times S^{d-1}} f(x_1) \cdots f(x_k) \mathbf{1}\{(x_{k+1}, \dots, x_{k+k'}) \in \cdot\} \Xi_{d-1,\dots,d-1,m}\Big(u_1^-, \dots \\ &\qquad \dots, u_k^-, K_1, \dots, K_{k'}; d(x_1, \dots, x_{k+k'}, u)\Big) \end{aligned}$$

defines a measure on $\left(\mathbb{R}^d\right)^{k'} \times S^{d-1}$. Furthermore, we define by

$$\Psi_m(u_1, \dots, u_k, K_1, \dots, K_{k'}; C) := \Xi_m\Big(u_1, \dots, u_k, K_1, \dots, K_{k'}; (\mathbb{R}^d)^{k'} \times C\Big)$$

special mixed measures on S^{d-1} and by

$$V_m(u_1, \dots, u_k, K_1, \dots, K_{k'}) := \Psi_m\Big(u_1, \dots, u_k, K_1, \dots, K_{k'}; S^{d-1}\Big)$$

the corresponding total measures.

In Definition 3.2.4 we have defined mixed support measures where the first k arguments are unit vectors and the following arguments are convex bodies. If this order is not given we define mixed measures in the following way. For $L_1, \ldots, L_k \in S^{d-1} \cup \mathcal{K}$ we define

$$\Xi_m(L_1, \ldots, L_k; \cdot) := \Xi_m(L_{\sigma(1)}, \ldots, L_{\sigma(k)}; \cdot),$$

where σ is the permutation which sorts $L_1, \ldots, L_k$ in such a way that the unit vectors occur first and the convex bodies afterwards but the order within the unit vectors respectively convex bodies remains.

In the following lemma we give the explicit representation of the just defined mixed measures in the case of polytopes. In particular we observe that the definition of the measures does not depend on the explicit choice of the function f as long as the integral property (3.1) is fulfilled with respect to the unit vectors which appear as arguments.

Lemma 3.2.5. *(a) Let $u_1, \ldots, u_k \in S^{d-1}$ and $K_1, \ldots, K_{k'}$ be polytopes. Then*

$$\begin{aligned}
&\Xi_m(u_1, \ldots, u_k, K_1, \ldots, K_{k'}; A_1 \times \ldots \times A_{k'} \times C) \\
&= 2^k \sum_{F_1 \in \mathcal{F}_{m_1}(K_1)} \cdots \sum_{F_{k'} \in \mathcal{F}_{m_{k'}}(K_{k'})} \frac{\mathcal{H}^{d-1-j}\left(\left(\sum_{i=1}^{k} \operatorname{pos}(u_i) + \sum_{i=1}^{k'} N(K_i, F_i)\right) \cap C\right)}{\omega_{d-j}} \\
&\times [u_1^\perp, \ldots, u_k^\perp, F_1, \ldots, F_{k'}](\mathcal{H}^{m_1} \llcorner F_1)(A_1) \cdots (\mathcal{H}^{m_{k'}} \llcorner F_{k'})(A_{k'}).
\end{aligned}$$

(b) The map

$$(u_1, \ldots, u_k, K_1, \ldots, K_{k'}) \mapsto \Xi_m(u_1, \ldots, u_k, K_1, \ldots, K_{k'}; \cdot)$$

is weakly continuous on $\left(S^{d-1}\right)^k \times \mathcal{K}^{k'}$.

Proof. The representation in (a) follows from Theorem 3.2.2, (v) and the integral property (3.1). Let $B \subset \mathbb{R}^d$ be some cube with $\operatorname{supp} f \subset B$. The mapping $u \mapsto u^- \cap B$ is continuous on S^{d-1}. Thus, Theorem 3.2.2, (vi) implies that the mapping

$$(u_1, \ldots, u_k, K_1, \ldots, K_{k'}) \mapsto \Xi_{d-1,\ldots,d-1,m}\left(u_1^- \cap B, \ldots, u_k^- \cap B, K_1, \ldots, K_{k'}; \cdot\right)$$

is weakly continuous on $(S^{d-1})^k \times \mathcal{K}^{k'}$. Together with the continuity of the function f this implies the assertion. □

The other properties of the mixed support measures collected in Theorem 3.2.2 have obvious extensions to the mixed measures of Definition 3.2.4.

Translative integral formula for mixed support measures

The mixed support measures fulfil on their part a translative integral formula, which follows from the homogeneity properties and a comparison of coefficients.

Lemma 3.2.6. *Let $K_1, \dots, K_{k+k'} \in \mathcal{K}$, $j \in \{0, \dots, d-1\}$, $j \le j' \le d$, $k, k' \in \mathbb{N}$ and $m = (m_1, \dots, m_k) \in \operatorname{mix}(d - j' + j, k)$, which implies $(m, j') \in \operatorname{mix}(j, k+1)$. Let $A_1, \dots, A_{k+k'} \subset \mathbb{R}^d$ and $C \subset S^{d-1}$ be Borel sets. Then*

$$\int_{(\mathbb{R}^d)^{k'-1}} \Xi_{m,j'}\Big(K_1, \dots, K_k, \Big(K_{k+1} \cap \bigcap_{i=2}^{k'} (K_{k+i} + x_i)\Big); A_1 \times \dots$$
$$\dots \times A_k \times \Big(A_{k+1} \cap \bigcap_{i=2}^{k'} (A_{k+i} + x_i)\Big) \times C\Big) d(x_2, \dots, x_{k'})$$
$$= \sum_{m' \in \operatorname{mix}(j',k')} \Xi_{m,m'}(K_1, \dots, K_{k+k'}; A_1 \times \dots \times A_{k+k'} \times C).$$

Proof. Observe that

$$\operatorname{mix}(j, k+1) = \{(m, j') : m \in \operatorname{mix}(d - j' + j, k), j \le j' \le d\}$$

and

$$\operatorname{mix}(j, k+k') = \{(m, m') : m \in \operatorname{mix}(d - j' + j, k), m' \in \operatorname{mix}(j', k'), j \le j' \le d\}.$$

Applying the translative integral formula for support measures, Theorem 3.2.1, to the convex bodies $K_1, \dots, K_k$ and $K_{k+1} \cap \bigcap_{i=2}^{k'} (K_{k+i} + x_{k+i} - x_{k+1})$ and using the translation invariance of the Lebesgue measure we get

$$\int_{(\mathbb{R}^d)^{k+k'-1}} \Xi_j\Big(K_1 \cap \bigcap_{i=2}^{k+k'} (K_i + x_i); A_1 \cap \bigcap_{i=2}^{k+k'} (A_i + x_i) \times C\Big) d(x_2, \dots, x_{k+k'})$$
$$= \sum_{j'=j}^{d} \sum_{m \in \operatorname{mix}(d-j'+j,k)} \int_{(\mathbb{R}^d)^{k'-1}} \Xi_{m,j'}\Big(K_1, \dots, K_k, K_{k+1} \cap \bigcap_{i=2}^{k'} (K_{k+i} + x_{k+i}); A_1 \times \dots$$
$$\dots \times A_k \times A_{k+1} \cap \bigcap_{i=2}^{k'} (A_{k+i} + x_{k+i}) \times C\Big) d(x_{k+2}, \dots, x_{k+k'}). \tag{3.2}$$

On the other hand an application of Theorem 3.2.1 to all convex bodies $K_1, \dots, K_{k+k'}$ leads to

$$\int_{(\mathbb{R}^d)^{k+k'-1}} \Xi_j\Big(K_1 \cap \bigcap_{i=2}^{k+k'} (K_i + x_i); A_1 \cap \bigcap_{i=2}^{k+k'} (A_i + x_i) \times C\Big) d(x_2, \dots, x_{k+k'})$$

$$= \sum_{j'=j}^{d} \sum_{m \in \mathrm{mix}(d-j'+j,k)} \sum_{m' \in \mathrm{mix}(j',k')} \Xi_{m,m'}(K_1,\ldots,K_{k+k'}; A_1 \times \ldots \times A_{k+k'} \times C). \quad (3.3)$$

Scaling (K_i, A_i) in (3.2) and (3.3) by $\lambda_i > 0$ for $1 \le i \le k$ and by $\lambda > 0$ for $i > k$ and using then the homogeneity properties of the mixed support measures and substituting x_i by λx_i for $i > k+1$ implies

$$\sum_{j'=j}^{d} \sum_{m \in \mathrm{mix}(d-j'+j,k)} \lambda_1^{m_1} \cdots \lambda_k^{m_k} \lambda^{(k'-1)d+j'}$$

$$\int_{(\mathbb{R}^d)^{k'-1}} \Xi_{m,j'}\Big(K_1,\ldots,K_k, \Big(K_{k+1} \cap \bigcap_{i=2}^{k'} (K_{k+i}+x_{k+i})\Big); A_1 \times \ldots$$

$$\ldots \times A_k \times \left(A_{k+1} \cap \bigcap_{i=2}^{k'} (A_{k+i}+x_{k+i})\right) \times C\Big) d(x_{k+2},\ldots,x_{k+k'})$$

$$= \sum_{j'=j}^{d} \sum_{m \in \mathrm{mix}(d-j'+j,k)} \lambda_1^{m_1} \cdots \lambda_k^{m_k} \lambda^{(k'-1)d+j'}$$

$$\sum_{m' \in \mathrm{mix}(j',k')} \Xi_{m,m'}(K_1,\ldots,K_{k+k'}; A_1 \times \ldots \times A_{k+k'} \times C),$$

which is a multivariate polynomial in $\lambda_1,\ldots,\lambda_k,\lambda$. Comparing the coefficients of the left- and right-hand side yields the assertion. □

3.3 Partial inversion of the translative integral formula

In this section we express sums of mixed support measures as linear combinations of translative integrals.

For $j \in \{0,\ldots,d-1\}$ we need repeatedly the collection of multi-indices

$$\mathrm{mix}(j) := \{(m_1,\ldots,m_l) \in \{j,\ldots,d-1\}^l : m_1+\ldots+m_l = (l-1)d+j, 1 \le l \le d-j\}.$$

For example it holds

$$\mathrm{mix}(d-1) = \{(d-1)\} \quad \text{and} \quad \mathrm{mix}(d-2) = \{(d-2),(d-1,d-1)\}.$$

For Borel sets $A_1,\ldots,A_k \subset \mathbb{R}^d$ and $C \subset S^{d-1}$, $k \in \mathbb{N}$ and $K_1,\ldots,K_k \in \mathcal{K}$ we define

$$T_j(K_1,\ldots,K_k; A_1,\ldots,A_k; C)$$

$$:= \int\limits_{(\mathbb{R}^d)^{k-1}} \Xi_j(K_1 \cap (K_2 + x_2) \cap \dots$$
$$\dots \cap (K_k + x_k); A_1 \cap (A_2 + x_2) \cap \dots \cap (A_k + x_k) \times C) d(x_2, \dots, x_k).$$

We write

$$T_j(K_1, \dots, K_k; C) := T_j\left(K_1, \dots, K_k; \left(\mathbb{R}^d\right)^k \times C\right)$$

and

$$T_j(K_1, \dots, K_k) := T_j\left(K_1, \dots, K_k; \left(\mathbb{R}^d\right)^k \times S^{d-1}\right).$$

Then, we obtain the following partial inversion formula.

Theorem 3.3.1. *Let* $[k] := \{1, \dots, k\}$*. Then*

$$\sum_{\substack{m \in \mathrm{mix}(j) \\ |m| = k}} \Xi_m(K_1, \dots, K_k; A_1 \times \dots \times A_k \times C)$$
$$= \sum_{l=1}^{k} (-1)^{k-l} \sum_{1 \le i_1 < \dots < i_l \le k} T_j(K_{i_1}, \dots, K_{i_l}; A_{i_1}, \dots, A_{i_l}; C)$$
$$\times \prod_{p \in \{1, \dots, k\} \setminus \{i_1, \dots, i_l\}} V_d(K_p \cap A_p).$$

Proof. For $l \in \{1, \dots, k\}$ let

$$B_l := \{(m_1, \dots, m_k) \in \mathrm{mix}(j, k) : m_l = d\}.$$

We define an additive function on the collection of subsets of $\mathrm{mix}(j, k)$ by

$$\varphi : \mathcal{P}(\mathrm{mix}(j, k)) \to \mathbb{R}, \quad B \mapsto \sum_{m \in B} \Xi_m(K_1, \dots, K_k; A_1 \times \dots \times A_k \times C).$$

By the translative integral formula Theorem 3.2.1 and by splitting $\mathrm{mix}(j, k)$ in the set of tuples $(m_1, \dots, m_k)$ with no entry equal to d and the set of those tuples with at least one entry equal to d, we obtain

$$T_j(K_1, \dots, K_k; A_1, \dots, A_k; C)$$
$$= \sum_{m \in \mathrm{mix}(j,k)} \Xi_m(K_1, \dots, K_k; A_1 \times \dots \times A_k \times C)$$
$$= \sum_{\substack{m \in \mathrm{mix}(j) \\ |m| = k}} \Xi_m(K_1, \dots, K_k; A_1 \times \dots \times A_k \times C) + \varphi\left(\bigcup_{p=1}^{k} B_p\right). \tag{3.4}$$

Since φ is additive, we can apply the inclusion-exclusion principle (see [KR97], p. 7), which yields

$$\begin{aligned}\varphi\left(\bigcup_{p=1}^{k} B_p\right) &= \sum_{l=1}^{k}(-1)^{l+1} \sum_{\{p_1,\dots,p_l\}\subset[k]} \varphi(B_{p_1}\cap\ldots\cap B_{p_l})\\ &= \sum_{l=0}^{k-1}(-1)^{k-l+1} \sum_{\substack{I\subset[k]\\|I|=l}} \varphi\left(\bigcap_{p\in[k]\setminus I} B_p\right).\end{aligned}$$

For each $l \in \{1,\dots,k\}$ the definition of φ, the symmetry and the decomposability property of the mixed support measures stated in Theorem 3.2.2, (i) and (ii) imply

$$\begin{aligned}\sum_{\substack{I\subset[k]\\|I|=l}} \varphi\left(\bigcap_{p\in[k]\setminus I} B_p\right) &= \sum_{\substack{I\subset[k]\\|I|=l}} \sum_{\substack{m\in\mathrm{mix}(j,k)\\ m_p=d, p\in[k]\setminus I}} \Xi_{m_1,\dots,m_k}(K_1,\dots,K_k; A_1\times\ldots\times A_k\times C)\\ &= \sum_{I=\{i_1,\dots,i_l\}\subset[k]} \prod_{p\in[k]\setminus I} V_d(K_p\cap A_p)\\ &\quad\times \sum_{m\in\mathrm{mix}(j,l)} \Xi_m(K_{i_1},\dots,K_{i_l}; A_{i_1}\times\ldots\times A_{i_l}\times C).\end{aligned}$$

Applying again the translative integral formula we get

$$\begin{aligned}\varphi\left(\bigcup_{p=1}^{k} B_p\right) &= \sum_{l=1}^{k-1}(-1)^{k-l+1} \sum_{I=\{i_1,\dots,i_l\}\subset[k]} T_j(K_{i_1},\dots,K_{i_l}; A_{i_1},\dots,A_{i_l}; C)\\ &\quad\times \prod_{p\in[k]\setminus I} V_d(K_p\cap A_p).\end{aligned}$$

Inserting this in (3.4) we obtain the assertion. □

Remark 3.3.2. We call the inversion formula partial because we have only a representation of a sum of mixed support measures. Still for the special case $k = d-j$ we obtain a complete inversion formula since on the left-hand side there occurs only the summand

$$\Xi_{d-1,\dots,d-1}(K_1,\dots,K_{d-j}; A_1\times\ldots\times A_{d-j}\times C).$$

3.4 Representation of special mixed support measures

In this subsection we show that some of the mixed support measures, namely those with some of the indices equal to $d-1$ can be represented as integrals with respect to products of the ordinary support measures Ξ_{d-1}. This representation is a main tool for the investigation of the mixed densities of geometric functionals derived from the support measures.

In [Wei01b, Theorem 9.1] Weil obtained a representation of special marginal measures of mixed curvature measures $\Phi_{d-1,\ldots,d-1,m'}$ with some of the indices equal to $d-1$ as an integral with respect to a product of $(d-1)$th area measures. Later, Goodey and Weil showed in [GW02, Theorem 1] that mixed curvature measures $\Phi_{d-1,\ldots,d-1,m'}$ have a representation as integrals with respect to a product of support measures Ξ_{d-1} where the integrand is a special mixed curvature measure with some of the convex body valued arguments replaced by halfspaces. This property complements the splitting property in the case where some of the indices are equal to d. The main ingredient of the proof is the explicit representation of the mixed curvature measures in the case of polytopes. Since an explicit representation of the mixed support measures for polytopes is also available (Theorem 3.2.1, (v)), Weil and Goodey's result can be extended to the setting of the mixed support measures.

For the formulation of the result we need the special mixed measures of Definition 3.2.4 where some of the convex body valued arguments are replaced by unit vectors. We obtain the following decomposition result.

Lemma 3.4.1. *Let* $j \in \{0,\ldots,d-1\}$, $k,k' \in \mathbb{N}$ *with* $1 \le k < d-j$ *and* $m \in \operatorname{mix}(k+j,k')$, $K_1,\ldots,K_{k+k'} \in \mathcal{K}$ *and let* $D := A_1 \times \ldots \times A_{k+k'} \times C \subset (\mathbb{R}^d)^{k+k'} \times S^{d-1}$ *be a Borel set. Then*

$$\begin{aligned}
&\Xi_{d-1,\ldots,d-1,m}(K_1,\ldots,K_{k+k'};D)\\
&= \int\limits_{\Sigma}\cdots\int\limits_{\Sigma}\prod_{i=1}^{k}\mathbf{1}_{A_i}(x_i)\Xi_m(u_1,\ldots,u_k,K_{k+1},\ldots,K_{k+k'};A_{k+1}\times\ldots\\
&\quad\ldots\times A_{k+k'}\times C)\Xi_{d-1}(K_1,d(x_1,u_1))\ldots\Xi_{d-1}(K_k,d(x_k,u_k)).
\end{aligned}$$

Proof. If $K_1,\ldots,K_{k+k'}$ are polytopes it follows from Theorem 3.2.2, (v) that

$$\begin{aligned}
&\Xi_{d-1,\ldots,d-1,m}(K_1,\ldots,K_{k+k'};A_1\times\ldots\times A_{k+k'}\times C)\\
&= \sum_{F_1\in\mathcal{F}_{d-1}(K_1)}\cdots\sum_{F_k\in\mathcal{F}_{d-1}(K_k)}\ \sum_{F_{k+1}\in\mathcal{F}_{m_1}(K_{k+1})}\cdots\sum_{F_{k+k'}\in\mathcal{F}_{m_{k'}}(K_{k+k'})}\\
&\quad\times\frac{\mathcal{H}^{d-1-j}\left(\left(\sum_{i=1}^{k+k'}N(K_i,F_i)\right)\cap C\right)}{\omega_{d-j}}[F_1,\ldots,F_{k+k'}](\mathcal{H}^{d-1}\llcorner F_1)(A_1)\cdots(\mathcal{H}^{d-1}\llcorner F_k)(A_k)\\
&\quad\times(\mathcal{H}^{m_1}\llcorner F_{k+1})(A_{k+1})\cdots(\mathcal{H}^{m_{k'}}\llcorner F_{k+k'})(A_{k+k'}).
\end{aligned}\tag{3.5}$$

We denote for a polytope K and $F \in \mathcal{F}_{d-1}(K)$ the outer unit normal of F by u_F. Then, we have $N(K,F) = \mathrm{pos}(u_F)$. By Lemma 3.1.2, (i) we have the identity

$$\Xi_{d-1}(K,\cdot) = \frac{1}{2} \sum_{F \in \mathcal{F}_{d-1}(K)} \delta_{u_F}\, \mathcal{H}^{d-1} \llcorner F,$$

where we use the Dirac measure $\delta_u(\cdot) := \mathbf{1}\{u \in \cdot\}, u \in S^{d-1}$. Furthermore, the linear subspace $u_{F_i}^{\perp}$ is parallel to the affine hull of F_i, which yields $[F_1,\dots,F_{k+k'}] = [u_{F_1}^{\perp},\dots,u_{F_k}^{\perp},F_{k+1},\dots,F_{k+k'}]$.

Using this in (3.5) yields

$$\begin{aligned}
&\Xi_{d-1,\dots,d-1,m}(K_1,\dots,K_{k+k'};A_1\times\dots\times A_{k+k'}\times C)\\
&= 2^k \int_{\Sigma^k} \sum_{F_{k+1}\in\mathcal{F}_{m_1}(K_{k+1})} \cdots \sum_{F_{k+k'}\in\mathcal{F}_{m_{k'}}(K_{k+k'})} \frac{\mathcal{H}^{d-1-j}\left(\left(\sum_{i=1}^{k}\mathrm{pos}(u_i) + \sum_{i=k+1}^{k+k'} N(K_i,F_i)\right)\cap C\right)}{\omega_{d-j}}\\
&\times [u_1^{\perp},\dots,u_k^{\perp},F_{k+1},\dots,F_{k+k'}] \prod_{i=1}^{k} \mathbf{1}_{A_i}(x_i)(\mathcal{H}^{m_1}\llcorner F_{k+1})(A_{k+1})\cdots(\mathcal{H}^{m_{k'}}\llcorner F_{k+k'})(A_{k+k'})\\
&\times \Xi_{d-1}(K_1,d(x_1,u_1))\dots\Xi_{d-1}(K_k,d(x_k,u_k)).
\end{aligned}$$

The assertion in the case of polytopes follows now since Lemma 3.2.5, (a) implies

$$\begin{aligned}
&\Xi_{m_1,\dots,m_{k'}}(u_1,\dots,u_k,K_{k+1},\dots,K_{k+k'};A_{k+1}\times\dots\times A_{k+k'}\times C)\\
&= 2^k \sum_{F_{k+1}\in\mathcal{F}_{m_1}(K_{k+1})} \cdots \sum_{F_{k+k'}\in\mathcal{F}_{m_{k'}}(K_{k+k'})} \frac{\mathcal{H}^{d-1-j}\left(\left(\sum_{i=1}^{k}\mathrm{pos}(u_i) + \sum_{i=k+1}^{k+k'} N(K_i,F_i)\right)\cap C\right)}{\omega_{d-j}}\\
&\times [u_1^{\perp},\dots,u_k^{\perp},F_{k+1},\dots,F_{k+k'}](\mathcal{H}^{m_1}\llcorner F_{k+1})(A_{k+1})\cdots(\mathcal{H}^{m_{k'}}\llcorner F_{k+k'})(A_{k+k'}).
\end{aligned}$$

For polytopes $K_1,\dots,K_{k+k'}$ and every continuous function $f : (\mathbb{R}^d)^{k+k'}\times S^{d-1}\to\mathbb{R}$ approximation by elementary functions implies that

$$\begin{aligned}
&\int_{(\mathbb{R}^d)^{k+k'}\times S^{d-1}} f(x_1,\dots,x_{k+k'},u)\Xi_{d-1,\dots,d-1,m}(K_1,\dots,K_{k+k'};d(x_1,\dots,x_k,u))\\
&= \int_{\Sigma}\dots\int_{\Sigma} f(x_1,\dots,x_{k+k'},u)\Xi_m(u_1,\dots,u_k,K_{k+1},\dots,K_{k+k'};d(x_{k+1},\dots,x_{k+k'},u)\\
&\quad \Xi_{d-1}(K_1,d(x_1,u_1))\dots\Xi_{d-1}(K_k,d(x_k,u_k)).
\end{aligned}$$

The weak continuity of the involved mixed measures implies the relation for arbitrary convex bodies $K_1,\dots,K_{k+k'}$ and continuous functions f. This proves the assertion. □

For $k \in \{1, \dots, n\}$ and $u_1, \dots, u_k \in S^{d-1}$ define a measure $\mu_k(u_1, \dots, u_k; \cdot)$ on S^{d-1} by

$$\mu_k(u_1, \dots, u_k; C) := \frac{2^k}{\omega_k} \nabla_k(u_1, \dots, u_k) \mathcal{H}^{k-1}(C \cap \operatorname{sconv}\{u_1, \dots, u_k\}), \tag{3.6}$$

where $C \in \mathcal{B}(S^{d-1})$. The mapping $(u_1, \dots, u_k) \mapsto \mu_k(u_1, \dots, u_k; \cdot)$ is weakly continuous on $(S^{d-1})^k$. Thus, the same arguments as in the proof of Lemma 3.4.1 lead to a representation of the mixed support measure with all indices equal to $d-1$.

Lemma 3.4.2. *Let $k \in \{1, \dots, d\}$, $K_1, \dots, K_k \in \mathcal{K}$ and let $D := A_1 \times \dots \times A_k \times C \subset (\mathbb{R}^d)^k \times S^{d-1}$ be a Borel set, then*

$$\begin{aligned} &\Xi_{d-1,\dots,d-1}(K_1, \dots, K_k; D) \\ &= \int_\Sigma \dots \int_\Sigma \mu_k(u_1, \dots, u_k; C) \prod_{i=1}^k \mathbf{1}_{A_i}(x_i)\, \Xi_{d-1}(K_1, d(x_1, u_1)) \dots \Xi_{d-1}(K_k, d(x_k, u_k)). \end{aligned}$$

The above relation can be interpreted as a special case of the representation of the mixed support measure obtained in Lemma 3.4.1 where $k' = 0$ if we define

$$\operatorname{mix}(k+j, 0) = \{\emptyset\}, \quad (d-1, \dots, d-1, \emptyset) := (d-1, \dots, d-1)$$

and

$$\Xi_\emptyset(u_1, \dots, u_k; C) := \mu_k(u_1, \dots, u_k; C), \quad u_1, \dots, u_k \in S^{d-1}.$$

As a consequence of Lemma 3.4.1 and Corollary 3.4.2 we obtain then the following representation for the mixed area measures with some of the indices equal to $d-1$.

Corollary 3.4.3. *Let $j \in \{0, \dots, d-1\}$, $k, k' \in \mathbb{N}$ with $1 \le k < d-j$ and $m \in \operatorname{mix}(k+j, k')$, $K_1, \dots, K_{k+k'} \in \mathcal{K}$ and $C \subset S^{d-1}$ be a Borel set. Then*

$$\begin{aligned} \Psi_{d-1,\dots,d-1,m}(K_1, \dots, K_{k+k'}; C) = \int_{S^{d-1}} \dots \int_{S^{d-1}} &\Psi_m(u_1, \dots, u_k, K_{k+1}, \dots, K_{k+k'}; C) \\ &\Psi_{d-1}(K_1, du_1) \dots \Psi_{d-1}(K_k, du_k) \end{aligned}$$

and in particular

$$\begin{aligned} &\Psi_{d-1,\dots,d-1}(K_1, \dots, K_k; C) \\ &= \int_{S^{d-1}} \dots \int_{S^{d-1}} \mu_k(u_1, \dots, u_k; C) \Psi_{d-1}(K_1, du_1) \dots \Psi_{d-1}(K_k, du_k). \end{aligned}$$

For special full mixed measures there exists a representation as integral with respect to the first area measure.

Lemma 3.4.4. *Let $K \in \mathcal{K}$ and $u \in S^{d-1}$. Then*

$$V_1(u,K) = \frac{2}{d-1} \int\limits_{S^{d-1}} \xi(u,v)\Psi_1(K,dv),$$

where $\xi(u,v) := g_d(-\langle u,v\rangle)$ with

$$g_2(t) := \frac{1}{\pi}(\pi - \arccos(t))\sqrt{1-t^2} - \frac{1}{2\pi}t, \quad t \in [-1,1],$$

$$g_3(t) := 1 + t\ln(1-t) + (\frac{4}{3} - \ln(2))t, \quad t \in (-1,1),$$

$$g_{k+2}(t) := \frac{k+1}{(k-1)^2}tg_k'(t) + \frac{k+1}{k-1}g_k(t) + \frac{k+1}{k+2}\frac{\omega_{k+1}}{\omega_{k+2}}t, \quad t \in (-1,1), k > 2.$$

Proof. In [SW08, Lemma 6.4.1] the relation

$$V_1(u,K) = 2h^*(K,-u)$$

to the centred support function h^* is shown. Berg [Ber69, (4)] derived the integral representation

$$h^*(K,u) = \frac{1}{d-1} \int\limits_{S^{d-1}} g_d(\langle u,v\rangle)\Psi_1(K,dv) \tag{3.7}$$

of the centred support function. The recursion formula for the functions g_d is derived in [Ber69, Theorem 3.3]. It is also shown that $g_d(\langle u,\cdot\rangle)$ is integrable on $(-1,1)$ for fixed $u \in S^{d-1}$. Observe that for $d \geq 3$ it is sufficient to define g_d on $(-1,1)$ since then $\Psi_1(K,\cdot)$ is an atom free measure for all $K \in \mathcal{K}$. This follows since by a result of Alexandrov [Ale96, Chapter V, §IV, Lemma I] the area measure, which is called curvature function in [Ale96], with j smaller than $d-1$ is atom free.

□

4 Density formulas for area measures of Boolean models

4.1 Mean value and density formulas

4.1.1 The isotropic situation

For an isotropic Boolean model the densities of the area measures evaluated at a fixed Borel set are proportional to the spherical Lebesgue measure by the following result.

Proposition 4.1.1. *Let Z be a stationary and isotropic Boolean model and $j \in \{0, \ldots, d-1\}$. Then, we have*

$$\overline{\Psi}_j(Z, C) = \omega_d^{-1} \overline{V}_j(Z)\, \sigma(C), \quad C \in \mathcal{B}(S^{d-1}).$$

Proof. For $r > 0$ we define a measure μ_r on S^{d-1} by

$$\mu_r(C) := \mathbb{E}\Psi_j \left(Z \cap rB^d, C\right)$$

for $C \in \mathcal{B}(S^{d-1})$.

The rotation covariance of Ψ_j and the isotropy of Z imply that the measure μ_r is rotation invariant. By their definition in Section 2.1 the positive part μ_r^+ and the negative part μ_r^- of μ_r inherit the rotation invariance from μ_r. Since μ_r^+ and μ_r^- are finite positive measures, they are both multiples of the Haar measure on S^{d-1}. Furthermore, the total measure is given by

$$\mu_r\left(S^{d-1}\right) = \mathbb{E}V_j(Z \cap rB^d).$$

Thus, we obtain

$$\mu_r\left(C\right) = \omega_d^{-1} \sigma(C)\, \mathbb{E}V_j\left(Z \cap rB^d\right), \quad C \in \mathcal{B}(S^{d-1}).$$

For fixed $C \in \mathcal{B}(S^{d-1})$ the functional $\Psi_j(\cdot, C)$ is geometric and by the definition (2.7) of the density of a Boolean model we have

$$\overline{\Psi}_j(Z, C) = \lim_{r \to \infty} \frac{1}{r^d \kappa_d} \mathbb{E}\Psi_j(Z \cap rB^d, C)$$

$$
\begin{aligned}
&= \lim_{r\to\infty} \frac{1}{r^d\kappa_d}\mu_r(C) \\
&= \lim_{r\to\infty} \frac{1}{r^d\kappa_d}\mathbb{E}V_j(Z\cap rB^d)\,\omega_d^{-1}\sigma(C) \\
&= \omega_d^{-1}\overline{V}_j(Z)\sigma(C)
\end{aligned}
$$

and thus the assertion. □

4.1.2 The non-isotropic situation

For a non-isotropic Boolean model the following density formulas for the scalar valued intrinsic volumes have been proven by Weil [Wei90, Corollary 7.5] and are also stated in [SW08, Theorem 9.1.5].

Theorem 4.1.2 (Weil 1990). *Let Z be a stationary Boolean model. Then*

$$
\overline{V}_d(Z) = 1 - \mathrm{e}^{-\overline{V}_d(X)} \quad \text{and} \quad \overline{V}_j(Z) = \mathrm{e}^{-\overline{V}_d(X)} \sum_{m\in\operatorname{mix}(j)} \frac{(-1)^{|m|-1}}{|m|!}\overline{V}_m(X,\dots,X)
$$

for $j = 0,\dots,d-1$.

The translative integral formula for support measures can be used to derive the following mean value formulas for support measures evaluated at a fixed Borel set.

Theorem 4.1.3. *Let Z be a stationary Boolean model, $W\in\mathcal{K}$, $j\in\{0,\dots,d-1\}$ and let $A\subset\mathbb{R}^d$ and $C\subset S^{d-1}$ be Borel sets.*

If $j=0$, then

$$
\begin{aligned}
&\mathbb{E}\left[\Xi_0(Z\cap W, A\times C)\right] \\
&= \Xi_0(W, A\times C) - \mathrm{e}^{-\overline{V}_d(X)} \sum_{m\in\operatorname{mix}(j)} \frac{(-1)^{|m|-1}}{|m|!}\Big[|m|\,\overline{\Xi}_m(W,X,\dots,X;A\times(\mathbb{R}^d)^{|m|-1}\times C) \\
&\qquad\qquad - \frac{1}{\omega_d}V_d(W\cap A)\,\overline{V}_m(X,\dots,X)\,\sigma(C)\Big]
\end{aligned}
$$

and if $j>0$, then

$$
\begin{aligned}
&\mathbb{E}\left[\Xi_j(Z\cap W, A\times C)\right] \\
&= \Xi_j(W, A\times C) - \mathrm{e}^{-\overline{V}_d(X)} \sum_{m\in\operatorname{mix}(j)} \frac{(-1)^{|m|-1}}{|m|!}\Big[|m|\,\overline{\Xi}_m(W,X,\dots,X;A\times(\mathbb{R}^d)^{|m|-1}\times C) \\
&\qquad\qquad - V_d(W\cap A)\,\overline{\Psi}_m(X,\dots,X;C)\Big].
\end{aligned}
$$

Proof. By (2.6), we have $\mathbb{E}[|\Xi_j(Z \cap W, A \times C)|] < \infty$ and

$$\mathbb{E}\left[\Xi_j(Z \cap W, A \times C)\right] = \sum_{k=1}^{\infty} \frac{(-1)^{k-1}}{k!}\gamma^k \int\limits_{\mathcal{K}_0} \dots \int\limits_{\mathcal{K}_0} T(W, K_1, \dots, K_k)\mathbb{Q}(dK_1)\dots\mathbb{Q}(dK_k),$$

where

$$T(W, K_1, \dots, K_k) = \int\limits_{(\mathbb{R}^d)^k} \Xi_j(W \cap (K_1 + x_1) \cap \dots \cap (K_k + x_k), A \times C)d(x_1, \dots, x_k).$$

Since the series in (2.6) converges absolutely, the function

$$T(W, \cdot, \dots, \cdot) : \mathcal{K}^k \to \mathbb{R}$$

is $\mathbb{Q}^k$-integrable. On $\mathcal{K}$ the mixed support measures evaluated at $A \times C$ are nonnegative, real-valued functionals and therefore they are also $\mathbb{Q}^k$-integrable by Theorem 3.2.1. This shows the existence of densities for the mixed support measures, though only in the case $A = \mathbb{R}^d$ the density can be formed with respect to the first argument. Since

$$\text{mix}(j, k+1) = \{(m_0, m) : m \in \text{mix}(d - m_0 + j, k), j \leq m_0 \leq d\},$$

an application of the iterated translative integral formula, Theorem 3.2.1 with $A_2 = \dots = A_k = \mathbb{R}^d$, yields

$$\mathbb{E}\left[\Xi_j(Z \cap W, A \times C)\right] = \sum_{k=1}^{\infty} \frac{(-1)^{k-1}}{k!}\gamma^k \sum_{m_0=j}^{d} \sum_{m \in \text{mix}(d-m_0+j,k)}^{d}$$
$$\int\limits_{\mathcal{K}_0^k} \Xi_{m_0,m}(W, K_1, \dots, K_k; A \times (\mathbb{R}^d)^k \times C)\mathbb{Q}^k(d(K_1, \dots, K_k)).$$

Introducing for $m = (m_1, \dots, m_k) \in \text{mix}(d - m_0 + j, k)$ the index l as the number of indices among $m_1, \dots, m_k$ that are smaller than d we obtain by the symmetry of the mixed support measures that

$$\mathbb{E}\left[\Xi_j(Z \cap W, A \times C)\right]$$
$$= \sum_{k=1}^{\infty} \frac{(-1)^{k-1}}{k!}\gamma^k \sum_{m_0=j}^{d} \sum_{l=\mathbf{1}\{m_0>j\}}^{(m_0-j)\wedge k} \binom{k}{k-l}$$
$$\times \sum_{\substack{m \in \text{mix}(d-m_0+j) \\ |m|=l}} \int\limits_{\mathcal{K}_0^k} \Xi_{m_0,m,\underbrace{d,\dots,d}_{k-l \text{ times}}}(W, K_1, \dots, K_k; A \times (\mathbb{R}^d)^k \times C)\mathbb{Q}^k(d(K_1, \dots, K_k)).$$

Then, we rearrange the summation using the index $p = k - l$ and use the decomposability of the mixed support measures, which yields

$$
\begin{aligned}
&\mathbb{E}\left[\Xi_j(Z \cap W, A \times C)\right] \\
&= \sum_{m_0=j}^{d} \sum_{l=\mathbf{1}\{m_0>j\}}^{m_0-j} \sum_{p=\mathbf{1}\{m_0=j\}}^{\infty} \frac{(-1)^{p+l-1}}{(p+l)!} \binom{p+l}{p} \\
&\quad \times \sum_{\substack{m \in \operatorname{mix}(d-m_0+j) \\ |m|=l}}^{d-1} \overline{\Xi}_{m_0,m}(W, X, \ldots, X; A \times (\mathbb{R}^d)^l \times C) \overline{V}_d(X)^p \\
&= \sum_{m_0=j+1}^{d} \sum_{l=1}^{m_0-j} \frac{(-1)^{l-1}}{l!} \sum_{\substack{m \in \operatorname{mix}(d-m_0+j) \\ |m|=l}} \overline{\Xi}_{m_0,m}(W, X, \ldots, X; A \times (\mathbb{R}^d)^l \times C) \mathrm{e}^{-\overline{V}_d(X)} \\
&\quad + \Xi_j(W, A \times C) \left[1 - \mathrm{e}^{-\overline{V}_d(X)}\right] \\
&= \sum_{m_0=j+1}^{d} \sum_{m \in \operatorname{mix}(d-m_0+j)} \frac{(-1)^{|m|-1}}{|m|!} \overline{\Xi}_{m_0,m}(W, X, \ldots, X; A \times (\mathbb{R}^d)^l \times C) \mathrm{e}^{-\overline{V}_d(X)} \\
&\quad + \Xi_j(W, A \times C) \left[1 - \mathrm{e}^{-\overline{V}_d(X)}\right]
\end{aligned}
$$

If we use the decomposability of the support measures in the case $m_0 = d$ and the relation

$$
\operatorname{mix}(j) = \{(m_0, m) : j+1 \leq m_0 \leq d-1, m \in \operatorname{mix}(d - m_0 + j)\} \cup \{(j)\}
$$

we obtain

$$
\begin{aligned}
&\mathbb{E}\left[\Xi_j(Z \cap W, A \times C)\right] \\
&= \Xi_j(W, A \times C) + \sum_{m \in \operatorname{mix}(j)} \frac{(-1)^{|m|}}{|m|!} \| m | \overline{\Xi}_m(W, X, \ldots, X; A \times (\mathbb{R}^d)^{|m|-1} \times C) \mathrm{e}^{-\overline{V}_d(X)} \\
&\quad - V_d(W \cap A) \sum_{m \in \operatorname{mix}(j)} \frac{(-1)^{|m|-1}}{|m|!} \overline{\Psi}_m(X, \ldots, X; C) \mathrm{e}^{-\overline{V}_d(X)},
\end{aligned}
$$

which yields the assertion for $1 \leq j \leq d-1$. A special situation occurs in the case $j = 0$. Then, the area measure Ψ_0 is proportional to the spherical Lebesgue measure.

As a consequence we can apply the translative integral formula Theorem 3.2.1 to the total measure $\Psi_0(\cdot, S^{d-1}) = V_0(\cdot)$ and to the measure $\Psi_0(\cdot, C)$ evaluated at some Borel set $C \subset S^{d-1}$ and compare the right-hand sides. Since the mixed measures are homogeneous with respect to each of the arguments $K_1, \ldots, K_k$ we can scale each convex body and compare the coefficients of the multivariate polynomial with respect to the scaling factors. This yields

$$\Psi_m(K_1,\dots,K_k;C)=\frac{1}{\omega_d}\,V_m(K_1,\dots,K_k)\,\sigma(C),\quad m\in\operatorname{mix}(0,k).$$

□

As a corollary of the above theorem we obtain a density formula for area measures evaluated at a fixed Borel set.

Corollary 4.1.4. *Let Z be a stationary Boolean model, $j\in\{0,\dots,d-1\}$ and $C\subset S^{d-1}$ be a Borel set.*

If $j=0$, then

$$\overline{\Psi}_0(Z,C)=\frac{1}{\omega_d}\,\mathrm{e}^{-\overline{V}_d(X)}\sum_{m\in\operatorname{mix}(j)}\frac{(-1)^{|m|-1}}{|m|!}\overline{V}_m(X,\dots,X)\,\sigma(C)$$

and if $j>0$, then

$$\overline{\Psi}_j(Z,C)=\mathrm{e}^{-\overline{V}_d(X)}\sum_{m\in\operatorname{mix}(j)}\frac{(-1)^{|m|-1}}{|m|!}\overline{\Psi}_m(X,\dots,X;C).$$

Proof. If we choose $A=\mathbb{R}^d$ in Theorem 4.1.3 we obtain

$$\begin{aligned}&\overline{\Psi}_j(Z,C)\\&=\lim_{r\to\infty}\frac{1}{V_d(rW)}\mathbb{E}\Psi_j(Z\cap rW,C)\\&=\lim_{r\to\infty}\frac{1}{V_d(rW)}\Big[\Psi_j(rW,C)\\&-e^{-\overline{V}_d(X)}\sum_{m\in\operatorname{mix}(j)}\frac{(-1)^{|m|-1}}{|m|!}\left[|m|\overline{\Psi}_m(rW,X,\dots,X;C)-V_d(rW)\overline{\Psi}_m(X,\dots,X;C)\right]\Big].\end{aligned}$$

The homogeneity properties of the (mixed) area measures imply

$$\Psi_j(rW,C)=r^j\,\Psi_j(W,C)$$

and

$$\overline{\Psi}_m(rW,X,\dots,X;C)=r^{m_1}\,\overline{\Psi}_m(W,X,\dots,X;C)$$

for $m=(m_1,\dots,m_k)\in\operatorname{mix}(j)$. Observe that $m_1<d$ by the definition of $\operatorname{mix}(j)$. Thus several summands converge to zero and it remains

$$\overline{\Psi}_j(Z,C)=e^{-\overline{V}_d(X)}\sum_{m\in\operatorname{mix}(j)}\frac{(-1)^{|m|-1}}{|m|!}\overline{\Psi}_m(X,\dots,X;C).$$

□

4.1.3 Density formulas for mixed area measures

The translative integral formula for mixed support measures Lemma 3.2.6 implies a translative integral formula for mixed area measures. By the same arguments as in the proof of Theorem 4.1.3 we can obtain the following density formulas for mixed area measures where some of the arguments are fixed convex bodies or half-spaces.

Theorem 4.1.5. *Let Z be a stationary Boolean model, $L_1, \dots, L_k \in \mathcal{R}$, $j \in \{0, \dots, d-1\}$, $j \leq j' \leq d-1$, $1 \leq k < d-j$, $m \in \operatorname{mix}(d-j'+j, k)$ and $C \subset S^{d-1}$ a Borel set. Then,*

$$\overline{\Psi}_{j',m}(Z, L_1, \dots, L_k; C) = \mathrm{e}^{-\overline{V}_d(X)} \sum_{m' \in \operatorname{mix}(j')} \frac{(-1)^{|m'|-1}}{|m'|!} \overline{\Psi}_{m',m}(X, \dots, X, L_1, \dots, L_k; C)$$

and for $u_1, \dots, u_k \in S^{d-1}$ we get

$$\overline{\Psi}_{k+j}(Z, u_1, \dots, u_k; C) = \mathrm{e}^{-\overline{V}_d(X)} \sum_{m' \in \operatorname{mix}(k+j)} \frac{(-1)^{|m'|-1}}{|m'|!} \overline{\Psi}_{m'}(X, \dots, X, u_1, \dots, u_k; C).$$

Remark 4.1.6. If we define densities $\overline{T}_j(X, \dots, X; C)$ of translative integrals of area measures evaluated at a Borel set $C \subset S^{d-1}$ in the usual way, the partial inversion formula Theorem 3.3.1 applied to the right-hand side of Corollary 4.1.4 yields a representation of $\overline{\Psi}_j(Z, C)$ as linear combination of densities $\overline{T}_j(X, \dots, X; C)$ of translative integrals with at most $d-j$ entries X.

In the following it will be useful to define also densities of measures. Let $W \in \mathcal{K}$ with $V_d(W) > 0$, $j \in \{0, \dots, d-1\}$, $k \in \mathbb{N}$ and $m \in \operatorname{mix}(j, k)$. Then, densities of (mixed) area measures are defined by

$$\overline{\Psi}_m(X, \dots, X; A) := \gamma^k \int_{\mathcal{K}_0^k} \Psi_m(K_1, \dots, K_k; A) \mathbb{Q}^k(d(K_1, \dots, K_k)), \quad A \in \mathcal{B}(S^{d-1})$$

and

$$\overline{\Psi}_j(Z, A) = \lim_{\varrho \to \infty} \frac{\mathbb{E}\Psi_j(Z \cap \varrho W; A)}{V_d(\varrho W)}, \quad A \in \mathcal{B}(S^{d-1}).$$

The monotone convergence theorem implies that $\overline{\Psi}_m(X, \dots, X; \cdot)$ defines a measure. By Corollary 4.1.4 $\overline{\Psi}_j(Z, \cdot)$ is a finite signed measure.

4.2 Inversion of density formulas for area measures

So far, an inversion of the density formulas for area measures Ψ_j is possible for $j = d-1$ since Theorem 4.1.2 and Theorem 4.1.4 imply

$$\overline{V}_d(X) = -\ln(1 - \overline{V}_d(Z)) \quad \text{and} \quad \overline{\Psi}_{d-1}(X, \cdot) = \frac{1}{1 - \overline{V}_d(Z)} \overline{\Psi}_{d-1}(Z, \cdot). \tag{4.1}$$

The inversion of the density formulas for $j < d-1$ is more difficult because densities of mixed area measures are involved. In this section we show that in the most relevant dimensions $d = 2$ and $d = 3$ all densities of the area measures of the particle process X can be expressed by the densities of the area measures of the Boolean model. This is a direct consequence of Lemma 4.2.1 and Lemma 4.2.2. In particular we obtain in Corollary 4.2.4 and Corollary 4.2.5 formulas expressing the intensity γ in two respectively three dimensions in terms of the densities of the area measures of the Boolean model Z. These formulas imply that the intensity γ is in two dimensions uniquely determined by $\overline{V}_2(Z), \overline{\Psi}_1(Z,\cdot)$ and $\overline{V}_0(Z)$. In three dimensions γ is uniquely determined by $\overline{V}_3(Z), \overline{\Psi}_2(Z,\cdot), \overline{\Psi}_1(Z,\cdot)$ and $\overline{V}_0(Z)$.

These uniqueness results are in the spirit of results derived by Weil involving the surface area measure and support functions [Wei99] or mixed volumes [Wei01a].

In practise it is a complicated problem to estimate a measure-valued quantity. Thus we consider in the next chapter applications of the formulas for area measures to real- and tensor-valued quantities derived from area measures.

At first we express the densities of the area measure of the particle process of order $d-2$ by densities of the area measures of the Boolean model of order $d-2$ and $d-1$.

For the statement of the result we need the measure $\mu_k(u_1,\dots,u_k;\cdot)$ on the unit sphere which was defined in (3.6) for $u_1,\dots,u_k \in S^{d-1}$ and $k \in \mathbb{N}$.

Lemma 4.2.1. *Let Z be a stationary Boolean model, $A \in \mathcal{B}(S^{d-1})$. Then*

$$\overline{\Psi}_{d-2}(X,A) = \frac{1}{1-\overline{V}_d(Z)}\overline{\Psi}_{d-2}(Z,A) + \frac{1}{2}\left(\frac{1}{1-\overline{V}_d(Z)}\right)^2 \int_{S^{d-1}}\int_{S^{d-1}} \mu_2(u,v;A)\overline{\Psi}_{d-1}(Z,du)\overline{\Psi}_{d-1}(Z,dv).$$

Proof. We have

$$\operatorname{mix}(d-2) = \{(d-2),(d-1,d-1)\}.$$

Thus, it follows by Corollary 4.1.4 that

$$\overline{\Psi}_{d-2}(Z,A) = e^{-\overline{V}_d(X)}\left[\overline{\Psi}_{d-2}(X,A) - \frac{1}{2}\overline{\Psi}_{d-1,d-1}(X,X;A)\right]. \tag{4.2}$$

It is well-known (compare Theorem 4.1.2) that

$$e^{-\overline{V}_d(X)} = 1-\overline{V}_d(Z). \tag{4.3}$$

From Corollary 3.4.3 we obtain

$$\overline{\Psi}_{d-1,d-1}(X,X;A) = \int_{S^{d-1}}\int_{S^{d-1}} \mu_2(u,v;A)\overline{\Psi}_{d-1}(X,du)\overline{\Psi}_{d-1}(X,dv). \tag{4.4}$$

By Corollary 4.1.4 we have

$$\overline{\Psi}_{d-1}(Z,A) = e^{-\overline{V}_d(X)}\overline{\Psi}_{d-1}(X,A). \tag{4.5}$$

Solving (4.2) for $\overline{\Psi}_{d-2}(X,A)$ and combining (4.3), (4.4) and (4.5) we obtain the assertion. □

The densities of the area measure of the particle process of order $d-3$ have also a representation by densities of the measures of the Boolean model. Though, we are not able to obtain an expression involving only densities of the area measures of the Boolean model instead also the density of the measure $\overline{\Psi}_{d-2}(Z,u;\cdot)$ depending on a unit vector $u \in S^{d-1}$ is involved.

Lemma 4.2.2. *Let $d \geq 3$ and Z be a stationary Boolean model and $A \in \mathcal{B}(S^{d-1})$. Then*

$$\begin{aligned}
&\overline{\Psi}_{d-3}(X,A) \\
&= \frac{1}{1-\overline{V}_d(Z)}\overline{\Psi}_{d-3}(Z,A) \\
&+ \left(\frac{1}{1-\overline{V}_d(Z)}\right)^2 \int_{S^{d-1}} \overline{\Psi}_{d-2}(Z,u;A)\overline{\Psi}_{d-1}(Z,du) \\
&+ \frac{1}{3}\left(\frac{1}{1-\overline{V}_d(Z)}\right)^3 \int_{S^{d-1}}\int_{S^{d-1}}\int_{S^{d-1}} \mu_3(u,v,w;A)\overline{\Psi}_{d-1}(Z,du)\overline{\Psi}_{d-1}(Z,dv)\overline{\Psi}_{d-1}(Z,dw).
\end{aligned}$$

Proof. Since

$$\operatorname{mix}(d-3) = \{(d-3),(d-1,d-2),(d-2,d-1),(d-1,d-1,d-1)\},$$

Corollary 4.1.4 yields

$$\begin{aligned}
&\overline{\Psi}_{d-3}(Z,A) \\
&= e^{-\overline{V}_d(X)}\left[\overline{\Psi}_{d-3}(X,A) - \overline{\Psi}_{d-2,d-1}(X,X;A) + \frac{1}{6}\overline{\Psi}_{d-1,d-1,d-1}(X,X,X;A)\right].
\end{aligned} \tag{4.6}$$

By Corollary 3.4.3, by the relation

$$\Psi_{d-2}(Z,u;A) = e^{-\overline{V}_d(X)}\left[\overline{\Psi}_{d-2}(X,u;A) - \frac{1}{2}\overline{\Psi}_{d-1,d-1}(X,X,u;A)\right], \quad u \in S^{d-1},$$

which is a special case of the second formula of Theorem 4.1.5, and by (4.3) we obtain

$$\begin{aligned}
&\overline{\Psi}_{d-2,d-1}(X,X;A) \\
&= \int_{S^{d-1}} \overline{\Psi}_{d-2}(X,u;A)\overline{\Psi}_{d-1}(X,du)
\end{aligned}$$

$$= \int_{S^{d-1}} \left[\frac{1}{1-\overline{V}_d(Z)} \overline{\Psi}_{d-2}(Z,u;A) + \frac{1}{2}\overline{\Psi}_{d-1,d-1}(X,X,u;A) \right] \overline{\Psi}_{d-1}(X,du).$$

Then, (4.5) and Corollary 3.4.3 imply

$$\overline{\Psi}_{d-2,d-1}(X,X;A) = \left(\frac{1}{1-\overline{V}_d(Z)}\right)^2 \int_{S^{d-1}} \overline{\Psi}_{d-2}(Z,u;A)\overline{\Psi}_{d-1}(Z,du) + \frac{1}{2}\overline{\Psi}_{d-1,d-1,d-1}(X,X,X;A) \tag{4.7}$$

and

$$\overline{\Psi}_{d-1,d-1,d-1}(X,X,X;A) = \left(\frac{1}{1-\overline{V}_d(Z)}\right)^3 \int_{S^{d-1}} \int_{S^{d-1}} \int_{S^{d-1}} \mu_3(u,v,w;A)\overline{\Psi}_{d-1}(Z,du) \overline{\Psi}_{d-1}(Z,dv)\overline{\Psi}_{d-1}(Z,dw). \tag{4.8}$$

Solving for $\overline{\Psi}_{d-3}(X,A)$ in (4.6) and combining (4.7) and (4.8) yields the assertion. □

Remark 4.2.3. We do not obtain an inversion formula for $\overline{\Psi}_{d-4}(X,\cdot)$. The reason is that we would have to express the densities of the area measures

$$\overline{\Psi}_m(X,\dots,X;\cdot), \quad m \in \operatorname{mix}(d-4) \setminus \{(d-4)\}$$

using densities of the area measures of the Boolean model Z. We have

$$\operatorname{mix}(d-4) = \Big\{(d-4),(d-3,d-1),(d-1,d-3),(d-2,d-2),(d-2,d-1,d-1), (d-1,d-2,d-1),(d-1,d-1,d-2),(d-1,d-1,d-1,d-1)\Big\}.$$

The densities of the area measures with multi-indices where only one of the indices is not equal to $d-1$ can be treated using Lemma 3.4.1 and the density formulas for area measures. But the densities of the area measures with the multi-index $(d-2,d-2)$ cannot be expressed in this way.

In two and three dimensions we obtain formulas for the intensity γ. Namely, if we choose $A=\mathbb{R}^d$ in Lemma 4.2.1 we obtain the following result.

Corollary 4.2.4. *Let Z be a stationary Boolean model in $\mathbb{R}^2$. Then*

$$\gamma = \overline{V}_0(X) = \frac{1}{1-\overline{V}_2(Z)}\overline{V}_0(Z) + \frac{1}{2}\left(\frac{1}{1-\overline{V}_2(Z)}\right)^2 \int_{S^1}\int_{S^1} \mu_2\left(u,v;S^1\right)\overline{\Psi}_1(Z,du)\overline{\Psi}_1(Z,dv).$$

For $d = 3$ we can apply Lemma 3.4.4 in Lemma 4.2.2, which yields the following formula.

Corollary 4.2.5. *Let Z be a stationary Boolean model in $\mathbb{R}^3$. Then*

$$\begin{aligned}\gamma = \overline{V}_0(X) &= \frac{1}{1-\overline{V}_3(Z)}\overline{V}_0(Z)\\ &+ \left(\frac{1}{1-\overline{V}_3(Z)}\right)^2 \int\limits_{S^2}\int\limits_{S^2} \xi(u,v)\overline{\Psi}_1(Z,du)\overline{\Psi}_2(Z,dv)\\ &+ \frac{1}{3}\left(\frac{1}{1-\overline{V}_3(Z)}\right)^3 \int\limits_{S^2}\int\limits_{S^2}\int\limits_{S^2} \mu_3(u,v,w;S^2)\overline{\Psi}_2(Z,du)\overline{\Psi}_2(Z,dv)\overline{\Psi}_2(Z,dw),\end{aligned}$$

where ξ is defined in Lemma 3.4.4.

Remark 4.2.6. We have derived the formulas for the intensity γ, which are stated in Corollary 4.2.4 and Corollary 4.2.5, as special cases of the formulas for $\overline{\Psi}_{d-2}(X,\cdot)$ and $\overline{\Psi}_{d-3}(X,\cdot)$ from Lemma 4.2.1 respectively Lemma 4.2.2. Alternatively, the formulas for the intensity can be obtained by combining the results in [Wei99] (or [SW08, p. 433 and p. 441]) with Berg's representation (3.7) of the centred support function.

5 Application to Minkowski tensors

In Chapter 3.1 we have derived mean value formulas for the support measures of a Boolean model observed in a window (Theorem 4.1.3) and we obtained formulas for the densities of the area measures of a Boolean model (Proposition 4.1.1 and Corollary 4.1.4). However, for the classification of Boolean models or the comparison of two Boolean models it is important to capture the significant properties with as little redundancy as possible. For this purposes it is desirable to study mean values of less-complicated functionals derived from the support measures.

For isotropic Boolean models the intrinsic volume densities have already shown to be a useful choice [AKM03, AKM09]. On the foundational side, the importance of the intrinsic volumes $V_0, \dots, V_d$ is expressed by Hadwiger's [Had57] famous characterization theorem, which states that the intrinsic volumes are a basis of the space of real-valued continuous, additive and motion invariant functionals on the space of convex bodies $\mathcal{K}$. As a consequence of the motion invariance, the intrinsic volumes reach their limits when it comes to the proper characterization of non-isotropic structures. Therefore one is interested in finding functionals which are sensitive to anisotropy and have as little redundancy as possible.

The results of Section 5.1 to 5.5 are published in the joint article [HHKM14] of the author with Daniel Hug, Michael Klatt and Klaus Mecke.

5.1 Minkowski tensors

In this section we introduce the Minkowski tensors, a collection of tensor-valued functionals derived from the support measures. Many important physical properties like elasticity, conductance and permeability are tensorial. Especially in physics the Minkowski tensors have recently become popular shape descriptors for anisotropic structures, see [STMK+11, STMK+13] and the literature cited therein.

We denote by $\mathbb{T}^p$ the vector space of symmetric tensors of rank p over $\mathbb{R}^d$. We use the scalar product to identify $\mathbb{R}^d$ with its dual space; then $\mathbb{T}^p$ can be viewed as the vector space of symmetric p-linear functionals on $\mathbb{R}^d$. A tensor $T \in \mathbb{T}^p$ is uniquely determined by the $\binom{d+p-1}{p}$ values $T_{i_1,\dots,i_p} := T(e_{i_1}, \dots, e_{i_p})$, $1 \le i_1 \le \dots \le i_p \le d$. Therefore, we can identify $\mathbb{T}^p$ with a $\binom{d+p-1}{p}$-dimensional Euclidean space, a fact which will be often useful. The symmetric tensor product $ab \in \mathbb{T}^{r+s}$ of symmetric tensors $a \in \mathbb{T}^r$ and $b \in \mathbb{T}^s$ for $r, s \in \mathbb{N}$ is defined by

$$(ab)_{i_1,\ldots,i_{r+s}} := \frac{1}{(r+s)!} \sum_{\pi \in S_{r+s}} a_{i_{\pi(1)}} \cdots a_{i_{\pi(r)}} b_{i_{\pi(r+1)}} \cdots b_{i_{\pi(r+s)}}, \quad 1 \leq i_1, \ldots, i_{r+s} \leq d,$$

where S_{r+s} denotes the symmetric group of order $r+s$. We write x^r for the r-fold symmetric tensor product of $x \in \mathbb{R}^d$. The metric tensor $Q \in \mathbb{T}^2$ is defined by $Q(x,y) = \langle x, y \rangle$, for $x, y \in \mathbb{R}^d$. The application of a rotation $g \in O_d$ to a tensor $T \in \mathbb{T}^p$ is defined by

$$(\vartheta T)_{i_1,\ldots,i_p} := T\left(g^{-1}e_{i_1}, \ldots, g^{-1}e_{i_p}\right), \quad 1 \leq i_1 \leq \ldots \leq i_p \leq d.$$

Originally the Minkowski tensors have been introduced in the context of convex geometric analysis where the characterization of additive functionals (valuations) on the space of convex bodies $\mathcal{K}$ enjoying specific properties is a highly active field of research. A valuation φ on $\mathcal{K}$ with values in the tensor space $\mathbb{T} := \bigoplus_{p=0}^{\infty} \mathbb{T}^p$ is called isometry covariant if

$$\varphi(gK) = g\varphi(K), \quad g \in O_d$$

and if φ has a polynomial behaviour with respect to translation of K. This means that there is some $s \in \mathbb{N}_0$ and there are valuations $\varphi_j : \mathcal{K} \to \bigoplus_{p=0}^{s} \mathbb{T}^p$ such that

$$\varphi(K+t) = \sum_{j=0}^{s} \varphi_j(K) t^{s-j}, \quad t \in \mathbb{R}^d, K \in \mathcal{K}.$$

Generalizations of Hadwiger's result, which concerns scalar-valued functionals, to vector-valued valuations which are isometry covariant have already been found in the early '70s by Hadwiger and Schneider [HS71, Sch72a, Sch72b].

More recently, tensor-valued valuations of higher rank have come into focus and it immediately turned out that in this case a basis cannot be determined that easily. The current mathematical study of tensor valuations has been initiated by McMullen [McM97]. For $K \in \mathcal{K}$, integers $r, s \geq 0$ and $0 \leq j \leq d-1$, the Minkowski tensors are defined by

$$\Phi_j^{r,s}(K) := c_{d-j}^{r,s} \int_{\Sigma} x^r u^s \Xi_j(K, d(x,u)) \tag{5.1}$$

and

$$\Phi_d^{r,0}(K) := \frac{1}{r!} \int_K x^r dx, \tag{5.2}$$

where $c_k^{r,s} := \frac{1}{r!s!} \frac{\omega_k}{\omega_{k+s}}$ for $1 \leq k \leq d$. From the properties of the support measures the following properties of the Minkowski tensors can be derived.

Lemma 5.1.1. *The Minkowski tensor $\Phi_j^{r,s}$ is positive homogeneous of degree $r+j$, continuous, additive and isometry covariant on $\mathcal{K}$. The isometry covariance is of the form*

$$\Phi_j^{r,s}(gK+t) = \sum_{k=0}^{r} \frac{1}{(r-k)!} t^{r-k}\, g\, \Phi_j^{k,s}(K), \quad t \in \mathbb{R}^d, g \in O_d, K \in \mathcal{K}.$$

In [McM97] McMullen conjectured that the *basic tensor valuations* $Q^m\Phi_j^{r,s}, r, s, m \in \mathbb{N}_0$ with $r+s+2m=p$, span the space of continuous, additive and isometry covariant $\mathbb{T}^p$-valued functionals, for every $p \in \mathbb{N}_0$. Furthermore, it was already observed by McMullen that the basic tensor valuations satisfy the linear dependencies

$$2\pi \sum_{s=0}^{\infty} s\Phi_{j-p+s}^{p-s,s} - Q \sum_{s=0}^{\infty} \Phi_{j-p+s}^{p-s,s-2} = 0, \quad j, p \in \mathbb{N}_0, \tag{5.3}$$

where $\Phi_j^{r,s} := 0$ if $j < 0$ or $r, s \notin \mathbb{N}_0$ or $r = d$ and $s > 0$. By (5.3) the basic tensor valuations do not form a basis of the vector space they span. McMullen's conjecture was almost immediately confirmed by Alesker [Ale99a, Ale99b]. In [HSS08b] it is shown that the linear dependencies (5.3) are up to linear combinations and multiplications by Q the only ones. In addition, it is shown how a basis can be constructed and the dimensions of the corresponding vector spaces are determined. In applications to Boolean models often only translation invariant functionals are considered. The space of continuous, additive, translation invariant and rotation covariant $\mathbb{T}^p$-valued functionals on $\mathcal{K}$ is spanned by the basic tensor valuations $Q^m\Phi_j^{0,s}, s, m \in \mathbb{N}_0$ with $s+2m=p$. The only linear dependencies (up to multiplication by Q and linear combinations) between the translation invariant basic tensor valuations are

$$2\pi s\Phi_0^{0,s} - Q\Phi_0^{0,s-2} = 0, \quad s \in \mathbb{N} \quad \text{and} \quad \Phi_0^1 = 0, \tag{5.4}$$

which is equivalent to

$$\Phi_0^{0,s} = \mathbf{1}\{s \in 2\mathbb{N}_0\} \frac{2}{s!\omega_{s+1}} Q^{s/2} V_0, \quad s \in \mathbb{N}_0. \tag{5.5}$$

More information on the mathematical and physical background of the Minkowski tensors can be found in [Sch00, HSS08a, HSS08b, STMK+11, STMK+13]. Of the many characterization theorems for valuations with values in some abelian group G and related to the present work, we only mention [Sch78, Lud02, Lud03, Lud13, Sch13b, HS14] which are concerned with characterizations of curvature measures, moment vectors, moment matrices, covariance matrices and local tensor valuations.

5.2 Mixed Minkowski tensors

We use Theorem 3.2.1 for the study of the translative integral of a Minkowski tensor. In the following we shall apply integrals and limits to tensors meaning the application to the real-valued coordinates. An important role is played by the following mixed tensorial functionals.

Let $K_1, \ldots, K_k \in \mathcal{R}$, $j \in \{0, \ldots, d-1\}$, $k \in \mathbb{N}$, $m \in \operatorname{mix}(j,k)$ and $r, s \in \mathbb{N}_0$. Then, we define mixed Minkowski tensors by

$$\Phi_m^{r,s}(K_1, \ldots, K_k) := c_{d-j}^{r,s} \int_{(\mathbb{R}^d)^k \times S^{d-1}} x_1^r u^s \, \Xi_m(K_1, \ldots, K_k; d(x_1, \ldots, x_k, u)).$$

A special case of the mixed Minkowski tensors are the mixed functionals of translative integral geometry

$$V_m = \Phi_m^{0,0}.$$

The translative integral formula for support measures Theorem 3.2.1 leads to the following translative integral formula for Minkowski tensors.

Corollary 5.2.1.

$$\int_{(\mathbb{R}^d)^{k-1}} \Phi_j^{r,s}(K_1 \cap (K_2 + x_2) \cap \ldots \cap (K_k + x_k)) \, d(x_2, \ldots, x_k) = \sum_{m \in \operatorname{mix}(j,k)} \Phi_m^{r,s}(K_1, \ldots, K_k).$$

The (mixed) Minkowski tensors inherit various properties from the support measures, which we collect in a corollary.

Corollary 5.2.2. (i) $\Phi_{m_1,\ldots,m_k}^{r,s}(K_1, \ldots, K_k)$ *is symmetric with respect to permutations of* $\{2, \ldots, k\}$. *For* $r = 0$ *it is even symmetric with respect to permutations of* $\{1, \ldots, k\}$;

(ii) $\Phi_{d,m_2,\ldots,m_k}^{r,s}(K_1, \ldots, K_k) = \Phi_d^{r,0}(K_1) \, \Phi_{m_2,\ldots,m_k}^{0,s}(K_2, \ldots, K_k)$

and

$$\Phi_{m_1,d,m_3,\ldots,m_k}^{r,s}(K_1, \ldots, K_k) = V_d(K_2) \, \Phi_{m_1,m_3,\ldots,m_k}^{r,s}(K_1, K_3, \ldots, K_k);$$

(iii) $\Phi_{m_1,\ldots,m_k}^{r,s}(K_1, \ldots, K_k)$ *is positively homogeneous of degree* $m_1 + r$ *with respect to* K_1 *and of degree* m_i *with respect to* K_i *for* $i \geq 2$;

(iv) *if* $K_1, \ldots, K_k$ *are polytopes, then*

$$\Phi_{m_1,\ldots,m_k}^{r,s}(K_1, \ldots, K_k) = \frac{1}{r!s!} \frac{1}{\omega_{d-j+s}} \sum_{F_1 \in \mathcal{F}_{m_1}(K_1)} \cdots \sum_{F_k \in \mathcal{F}_{m_k}(K_k)} \int_{\left(\sum_{i=1}^k N(K_i, F_i)\right) \cap S^{d-1}} u^s \, \mathcal{H}^{d-1-j}(du)$$

$$\times [F_1, \dots, F_k] \int_{F_1} x_1^r \mathcal{H}^{m_1}(dx_1)\, \mathcal{H}^{m_2}(F_2) \cdots \mathcal{H}^{m_k}(F_k);$$

(v) *the map* $(K_1, \dots, K_k) \mapsto \Phi_m^{r,s}(K_1, \dots, K_k)$ *is additive and continuous on* $\mathcal{K}^k$ *and measurable on* $\mathcal{R}^k$.

5.3 Mean value and density formulas for Minkowski tensors

In this section we first establish connections between mean values of the Minkowski tensors of the intersection of Z with a compact, convex window W and the densities of the particle process X. For the translation invariant Minkowski tensors, we obtain thus in a second step corresponding relations between the densities of the Minkowski tensors of the Boolean model and the densities of the Minkowski tensors of the underlying particle process. As the special case $r = s = 0$ one obtains Weil's well-known formulas for the densities of the intrinsic volumes of a non-isotropic Boolean model.

5.3.1 The isotropic situation

Under the assumption of isotropy the densities of the Minkowski tensors of a Boolean model are proportional to the densities of the intrinsic volumes.

Proposition 5.3.1. *Let Z be a stationary and isotropic Boolean model, $j \in \{0, \dots, d-1\}$ and $s \in \mathbb{N}_0$. Then*

$$\overline{\Phi}_j^{0,s}(Z) = \mathbf{1}\{s \in 2\mathbb{N}_0\} \alpha_{d,j,s} Q^{\frac{s}{2}} \overline{V}_j(Z),$$

where

$$\alpha_{d,j,s} := \frac{2}{s!} \frac{\omega_{d-j}\, \omega_{s+d}}{\omega_d\, \omega_{d-j+s}\, \omega_{s+1}}.$$

Proof. By Proposition 4.1.1 we have

$$\overline{\Psi}_j(Z, \cdot) = \omega_d^{-1} \overline{V}_j(Z)\, \sigma(\cdot).$$

Thus, we get

$$\overline{\Phi}_j^{0,s}(Z) = \frac{1}{s!} \frac{\omega_{d-j}}{\omega_{d-j+s}} \int_{S^{d-1}} u^s\, \overline{\Psi}_j(Z, du) = \frac{1}{s!} \frac{\omega_{d-j}}{\omega_{d-j+s}\, \omega_d} \overline{V}_j(Z) \int_{S^{d-1}} u^s\, \sigma(du).$$

In [SS02, (24)] it is shown that

$$\int_{S^{d-1}} u^s \sigma(du) = \mathbf{1}\{s \in 2\mathbb{N}_0\} 2 \frac{\omega_{s+d}}{\omega_{s+1}} Q^{\frac{s}{2}},$$

which yields the assertion. □

5.3.2 The non-isotropic situation

For a non-isotropic Boolean model we obtain the following mean value formulas for Minkowski tensors as a corollary of Theorem 4.1.3

Corollary 5.3.2. *Let Z be a stationary Boolean model, $W \in \mathcal{K}$ and $r, s \in \mathbb{N}_0$. If $j = 0$, then*

$$\begin{aligned}\mathbb{E}\left[\Phi_0^{r,s}(Z \cap W)\right] = \Phi_0^{r,s}(W) - \mathrm{e}^{-\overline{V}_d(X)} \sum_{m \in \operatorname{mix}(j)} \frac{(-1)^{|m|-1}}{|m|!} \Big[& |m| \overline{\Phi}_m^{r,s}(W, X, \dots, X) \\ & - \mathbf{1}\{s \in 2\mathbb{N}_0\}\, \alpha_{d,0,s}\, Q^{\frac{s}{2}} \Phi_d^{r,0}(W) \overline{V}_m(X, \dots, X)\Big]\end{aligned}$$

and if $1 \le j \le d-1$, then

$$\begin{aligned}\mathbb{E}\left[\Phi_j^{r,s}(Z \cap W)\right] = \Phi_j^{r,s}(W) - \mathrm{e}^{-\overline{V}_d(X)} \sum_{m \in \operatorname{mix}(j)} \frac{(-1)^{|m|-1}}{|m|!} \Big[& |m| \overline{\Phi}_m^{r,s}(W, X, \dots, X) \\ & - \Phi_d^{r,0}(W) \overline{\Phi}_m^{0,s}(X, \dots, X)\Big]\end{aligned}$$

and

$$\mathbb{E}\left[\Phi_d^{r,0}(Z \cap W)\right] = \Phi_d^{r,0}(W) \left(1 - \mathrm{e}^{-\overline{V}_d(X)}\right).$$

We obtain the following Minkowski tensor density formulas in general dimension.

Corollary 5.3.3. *Let Z be a stationary Boolean model in $\mathbb{R}^d$ and $s \in \mathbb{N}_0$. If $j = 0$, then*

$$\begin{aligned}\overline{\Phi}_0^{0,s}(Z) &= \mathrm{e}^{-\overline{V}_d(X)} \mathbf{1}\{s \in 2\mathbb{N}_0\} c_{d-1}^{0,s} Q^{\frac{s}{2}} \sum_{m \in \operatorname{mix}(0)} \frac{(-1)^{|m|-1}}{|m|!} \overline{V}_m(X, \dots, X) \\ &= \mathbf{1}\{s \in 2\mathbb{N}_0\}\, \alpha_{d,0,s}\, Q^{\frac{s}{2}} \overline{V}_0(Z).\end{aligned}$$

and if $1 \le j \le d-1$, then

$$\overline{\Phi}_j^{0,s}(Z) = \mathrm{e}^{-\overline{V}_d(X)} \sum_{m \in \operatorname{mix}(j)} \frac{(-1)^{|m|-1}}{|m|!} \overline{\Phi}_m^{0,s}(X, \dots, X).$$

Remark 5.3.4. For an isotropic and stationary Boolean model the same arguments as in the proof of Proposition 5.3.1 can be used to show that

$$\overline{\Phi}_m^{0,s}(X,\dots,X) = \mathbf{1}\{s \in 2\mathbb{N}_0\}\alpha_{d,j,s}\overline{V}_m(X,\dots,X),$$

for $m \in \operatorname{mix}(j), s \in \mathbb{N}_0, j \in \{0,\dots,d-1\}$.

Thus, a comparison with the density formulas for intrinsic volumes from Theorem 4.1.2 shows that Corollary 5.3.3 coincides for an isotropic and stationary Boolean model with Proposition 5.3.1.

Remark 5.3.5. A comparison of the previous Corollary 5.3.3 with Theorem 4.1.2 shows that in the case $j = 0$ the Minkowski tensor densities do not contain more information than the scalar valued densities $\overline{V}_0(Z)$. Though we would like to point out that this is indeed not the case for the corresponding mean value formulas for finite section window W, compare Corollary 5.3.2 in the case $j = 0$. Namely, if for pairwise distinct $\varrho_0,\dots,\varrho_d > 0$ the mean values $\mathbb{E}\,[\Phi_0^{r,s}(Z \cap \varrho_k W)]$, for $k = 0,\dots,d$, are known, we can separate the summands of different homogeneity degree in the right-hand side of the corresponding equations by merely solving a system of linear equations. In particular, if additionally the density $\overline{V}_d(X)$ is known, we obtain the density $\overline{\Phi}_{1,1}^{r,s}(W,X)$ in the case $d = 2$ and the density $\overline{\Phi}_{1,2}^{r,s}(W,X)$ in the case $d = 3$.

5.3.3 Density formulas for mixed Minkowski tensors

We define mixed Minkowski tensors depending on unit vectors, which are derived from the special mixed area measures of Definition 3.2.4.

Let $k \in \mathbb{N}$, $j \in \{0,\dots,d-1\}$, $1 \le k' \le d-j$, $u_1,\dots,u_{k'} \in S^{d-1}$, $K_1,\dots,K_k \in \mathcal{K}$, $m \in \operatorname{mix}(k'+j,k)$ and $s \in \mathbb{N}_0$. Then we define

$$\Phi_m^{0,s}(K_1,\dots,K_k,u_1,\dots,u_{k'}) := 2^{k'} c_{d-j}^{0,s} \int\limits_{S^{d-1}} v^s \Psi_m\Big(K_1,\dots,K_k,u_1,\dots,u_{k'};dv\Big).$$

As before we use the abbreviation $V_m := \Phi_m^{0,0}$. Then the following density formulas for the mixed Minkowski tensors are an immediate consequence of Theorem 4.1.5.

Corollary 5.3.6. *Let Z be a stationary Boolean model in $\mathbb{R}^d$, $L_1,\dots,L_k \in \mathcal{R}$, $j \in \{0,\dots,d-1\}$, $j \le j' \le d-1$, $1 \le k \le d-j$, $m \in \operatorname{mix}(d-j'+j,k)$ and $s \in \mathbb{N}_0$. Then,*

$$\overline{\Phi}_{j',m}^{0,s}(Z,L_1,\dots,L_k) = \mathrm{e}^{-\overline{V}_d(X)} \sum_{m' \in \operatorname{mix}(j')} \frac{(-1)^{|m'|-1}}{|m'|!} \overline{\Phi}_{m',m}^{0,s}(X,\dots,X,L_1,\dots,L_k)$$

and for $u_1,\dots,u_k \in S^{d-1}$ we get

$$\overline{\Phi}_{k+j,m}^{0,s}(Z,u_1,\dots,u_k) = \mathrm{e}^{-\overline{V}_d(X)} \sum_{m' \in \operatorname{mix}(k+j)} \frac{(-1)^{|m'|-1}}{|m'|!} \overline{\Phi}_{m',m}^{0,s}(X,\dots,X,u_1,\dots,u_k).$$

5.4 Application to a parametric Boolean model in the plane

In this subsection we apply the formulas from Corollary 5.3.3 to a parametric class of planar Boolean models with ellipse particles studied in [STMK+11, Section 2.2]. We shall see that for this simple parametric model the obtained density formulas allow to extract useful information from observations of the Boolean model.

For $\alpha \in [0,\infty], \gamma > 0$ and $E \in \mathcal{K}_0$, let $Z_{\alpha,\gamma,E}$ be a stationary Boolean model with intensity γ and the grains obtained by rotating E by a random angle $\theta \in [0, 2\pi)$. For $\alpha < \infty$ we assume that the random angle θ has the probability density

$$f_\alpha(\theta) = c(\alpha)\,|\cos\theta|^\alpha, \qquad \text{for } \theta \in [0, 2\pi),$$

with

$$c(\alpha) := \frac{\Gamma(1+\frac{\alpha}{2})}{2\sqrt{\pi}\Gamma(\frac{\alpha+1}{2})},$$

that is, the grain distribution of $Z_{\alpha,\gamma,E}$ is

$$\mathbb{Q}(\cdot) = \int_0^{2\pi} \mathbf{1}\{\vartheta(\theta)E \in \cdot\} f_\alpha(\theta)\, d\theta,$$

where $\vartheta(\theta) \in SO(2)$ is the rotation by the angle θ. The grain distribution of $Z_{\infty,\gamma,E}$ is $\mathbb{Q} = \delta_E$. In the following, we call E the base grain and α the orientation parameter of the Boolean model. We specify in this particular case the formulas for the densities obtained in Corollary 5.3.3. For this we have to determine $\overline{\Phi}_1^{0,s}(X), s \in \mathbb{N}_0, \overline{V}_0(X)$ and $\overline{V}_{1,1}(X,X)$. Starting with the density of the surface tensor we obtain for $s \in \mathbb{N}_0$ that

$$\overline{\Phi}_1^{0,s}(X) = \gamma \int_{\mathcal{K}_0} \Phi_1^{0,s}(K)\mathbb{Q}(dK) = \gamma \int_0^{2\pi} \Phi_1^{0,s}(\vartheta(\theta)E) f_\alpha(\theta) d\theta.$$

In the following we identify a p-tensor with an element of $\mathbb{R}^{d^p}$ in the usual way. We obtain by [STMK+13, (8)], for $s \in \mathbb{N}_0$ and $i_1, \dots, i_s \in \{1,2\}$, that

$$\left(\Phi_1^{0,s}(\vartheta(\theta)E)\right)_{i_1,\dots,i_s} = \sum_{j_1,\dots,j_s=1}^{2} (\vartheta(\theta))_{i_1,j_1} \cdots (\vartheta(\theta))_{i_s,j_s} \left(\Phi_1^{0,s}(E)\right)_{j_1,\dots,j_s}$$

and therefore, for $0 \le l \le s$, that

$$\left(\overline{\Phi}_1^{0,s}(X)\right)_{\underbrace{1,\dots,1}_{l \text{ times}}\underbrace{2,\dots,2}_{s-l \text{ times}}}$$

$$= \gamma \sum_{j_1,\dots,j_s=1}^{2} \int_0^{2\pi} (\vartheta(\theta))_{1,j_1} \cdots (\vartheta(\theta))_{1,j_l} \vartheta(\theta))_{2,j_{l+1}} \cdots (\vartheta(\theta))_{2,j_s} f_\alpha(\theta) d\theta$$
$$\times \left(\Phi_1^{0,s}(E)\right)_{j_1,\dots,j_s}$$
$$= \mathbf{1}\{s \text{ even}\}\, \gamma \sum_{\substack{j=0 \\ j+l \text{ even}}}^{s} \sum_{k=0\vee(j-s+l)}^{j\wedge l} (-1)^{l-k} \binom{l}{k} \binom{s-l}{j-k} \prod_{m=1}^{\frac{l+j-2k}{2}} (2m-1)$$
$$\times \prod_{m=1}^{\frac{s-l-j+2k}{2}} (\alpha+2m-1) \prod_{m=1}^{s/2} (\alpha+2m)^{-1} \left(\Phi_1^{0,s}(E)\right)_{\underbrace{1,\dots,1}_{j \text{ times}}\underbrace{2,\dots,2}_{s-j \text{ times}}}, \tag{5.6}$$

since we obtain for the integral prefactor in the second line of the above equation for $s_1, \dots, s_4, s \in \mathbb{N}_0$ with $s_1 + \dots + s_4 = s$ that

$$\int_0^{2\pi} (\vartheta(\theta))_{1,1}^{s_1} (\vartheta(\theta))_{1,2}^{s_2} (\vartheta(\theta))_{2,1}^{s_3} (\vartheta(\theta))_{2,2}^{s_4} f_\alpha(\theta) d\theta$$
$$= c(\alpha) \int_0^{2\pi} (\cos\theta)^{s_1+s_4} |\cos\theta|^\alpha (-\sin\theta)^{s_2} (\sin\theta)^{s_3} d\theta$$
$$= \begin{cases} 0, & \text{if } s_1+s_4 \text{ or } s_2+s_3 \text{ is odd}, \\ (-1)^{s_2} \dfrac{\prod_{m=1}^{(s_2+s_3)/2} (2m-1) \prod_{m=1}^{(s_1+s_4)/2} (\alpha+2m-1)}{\prod_{m=1}^{s/2} (\alpha+2m)}, & \text{otherwise}, \end{cases}$$

by the symmetry properties of sine and cosine and since

$$\int_0^{\frac{\pi}{2}} (\sin\varphi)^a (\cos\varphi)^b d\varphi = \frac{1}{2} \frac{\Gamma(\frac{a+1}{2})\Gamma(\frac{b+1}{2})}{\Gamma(\frac{a+b+2}{2})},$$

for $a, b > -1$; see [Art64, (5.6)] or [WW96, (12.42)]. In the case $s = 2$ equation (5.6) simplifies to

$$\overline{\Phi}_1^{0,2}(X) = \frac{\gamma}{\alpha+2} \begin{pmatrix} (\alpha+1)\left(\Phi_1^{0,2}(E)\right)_{1,1} + \left(\Phi_1^{0,2}(E)\right)_{2,2} & \alpha\left(\Phi_1^{0,2}(E)\right)_{1,2} \\ \alpha\left(\Phi_1^{0,2}(E)\right)_{1,2} & \left(\Phi_1^{0,2}(E)\right)_{1,1} + (\alpha+1)\left(\Phi_1^{0,2}(E)\right)_{2,2} \end{pmatrix}. \tag{5.7}$$

On the other hand, we obtain for the mixed density

$$\overline{V}_{1,1}(X,X) = \gamma^2 \int_0^{2\pi} \int_0^{2\pi} V_{1,1}(\vartheta(\theta_1)E, \vartheta(\theta_2)E) f_\alpha(\theta_1) f_\alpha(\theta_2) d\theta_1 d\theta_2$$

$$= \gamma^2 \int_0^{2\pi}\int_0^{2\pi} V_{1,1}(\vartheta(\theta_1-\theta_2)E,E) f_\alpha(\theta_1) f_\alpha(\theta_2) d\theta_1 d\theta_2$$

$$= \gamma^2 c(\alpha)^2 \int_0^{2\pi}\int_0^{2\pi} V_{1,1}(\vartheta(\theta_1)E,E)\, |\cos(\theta_1+\theta_2)|^\alpha |\cos(\theta_2)|^\alpha \, d\theta_1 d\theta_2, \tag{5.8}$$

where we have used that $V_{1,1}$ is invariant with respect to simultaneous rotations of its arguments and that the integrand is 2π-periodic with respect to θ_1.

Furthermore it follows from [Wei01b, Corollary 9.2] and the rotation covariance of the support measures for $\theta \in [0, 2\pi]$ that

$$V_{1,1}(\vartheta(\theta)E,E) = \frac{2}{\pi} \int_{\mathbb{R}^2\times S^1}\int_{\mathbb{R}^2\times S^1} \alpha\left(\vartheta(\theta)u_1,u_2\right) \sin\left(\alpha\left(\vartheta(\theta)u_1,u_2\right)\right)$$
$$\Xi_1(E, d(x_1,u_1))\Xi_1(E, d(x_2,u_2)), \tag{5.9}$$

where $\alpha(u_1,u_2) \in [0,\pi]$ denotes the smaller angle between $u_1, u_2 \in S^1$.

Remark 5.4.1. Assume that the above parametric Boolean model is observed and the densities

$$\overline{\Phi}_1^{0,2}(Z_{\alpha,\gamma,E}) \text{ and } \overline{V}_2(Z_{\alpha,\gamma,E})$$

are therefore known. Is it possible to obtain the parameters α and γ from the above densities of the Boolean model? To see that this is indeed the case, we use Corollary 5.3.3 to obtain

$$\overline{\Phi}_1^{0,2}(Z_{\alpha,\gamma,E}) = \overline{\Phi}_1^{0,2}(X)\, \mathrm{e}^{-\overline{V}_2(X)} \tag{5.10}$$

and (by Theorem 4.1.2)

$$\overline{V}_2(Z_{\alpha,\gamma,E}) = 1 - \mathrm{e}^{-\overline{V}_2(X)} = 1 - \mathrm{e}^{-\gamma V_2(E)}. \tag{5.11}$$

Thus, (5.11) yields

$$\gamma = -\frac{\ln\left(1-\overline{V}_2(Z_{\alpha,\gamma,E})\right)}{V_2(E)} \tag{5.12}$$

and, by (5.7) and (5.10),

$$\alpha = \frac{\gamma\left(\left(\Phi_1^{0,2}(E)\right)_{1,1} + \left(\Phi_1^{0,2}(E)\right)_{2,2}\right) - 2\,\mathrm{e}^{\gamma V_2(E)}\left(\overline{\Phi}_1^{0,2}(Z_{\alpha,\gamma,E})\right)_{1,1}}{\mathrm{e}^{\gamma V_2(E)}\left(\overline{\Phi}_1^{0,2}(Z_{\alpha,\gamma,E})\right)_{1,1} - \gamma\left(\Phi_1^{0,2}(E)\right)_{1,1}}. \tag{5.13}$$

In [HHKM14, Section 7.2] equations (5.12) and (5.13) are used to define estimators for the intensity γ and the orientation parameter α and the performance of the latter is tested in a simulation study.

Remark 5.4.2. In [STMK+11, Section 2.2] the Boolean model $Z_{\alpha,\gamma,E}$ with the base grain E being an ellipse is considered. Pixelized realizations of $[0,1]^2 \cap Z_{\alpha,\gamma,E}$ are used as input for testing the performance of real-valued characteristics derived from Minkowski tensors. More precisely, a so-called anisotropy index $\beta_1^{*\,0,2}$ is introduced, which is defined by

$$\beta_1^{*\,0,2} := \frac{\left(\Phi_1^{0,2}\left(Z_{\alpha,\gamma,E},[0,1]^2\right)\right)_{1,1}}{\left(\Phi_1^{0,2}\left(Z_{\alpha,\gamma,E},[0,1]^2\right)\right)_{2,2}},$$

where

$$\Phi_1^{0,2}\left(Z_{\alpha,\gamma,E},[0,1]^2\right) = c_1^{0,2} \int_{\mathbb{R}^2\times S^1} \mathbf{1}_{[0,1]^2}(x)\, u^2\, \Xi_1(Z_{\alpha,\gamma,E}, d(x,u)).$$

In [STMK+11, Section 2.2], $\langle\beta_1^{*\,0,2}\rangle$ denotes the mean value obtained by averaging $\beta_1^{*\,0,2}$ over several realizations of $Z_{\alpha,\gamma,E}$ and it is observed that for $\alpha = 0$, that is, in the isotropic case, we have $\langle\beta_1^{*\,0,2}\rangle = 1$. Furthermore, $\langle\beta_1^{*\,0,2}\rangle$ seems to be constant as function of the volume fraction $\overline{V}_2(Z_{\alpha,\gamma,E})$. Unfortunately, we are right now not able to explain these observations. But if instead of taking the mean value of $\beta_1^{*\,0,2}$, the mean value is taken separately for the denominator and nominator, that is, if

$$\frac{\left\langle\left(\Phi_1^{0,2}\left(Z_{\alpha,\gamma,E},[0,1]^2\right)\right)_{1,1}\right\rangle}{\left\langle\left(\Phi_1^{0,2}\left(Z_{\alpha,\gamma,E},[0,1]^2\right)\right)_{2,2}\right\rangle} \tag{5.14}$$

is considered, our previous results can be used to obtain some insight. The quantity (5.14) can be considered as an estimator of

$$\frac{\mathbb{E}\left[\left(\Phi_1^{0,2}\left(Z_{\alpha,\gamma,E},[0,1]^2\right)\right)_{1,1}\right]}{\mathbb{E}\left[\left(\Phi_1^{0,2}\left(Z_{\alpha,\gamma,E},[0,1]^2\right)\right)_{2,2}\right]}. \tag{5.15}$$

Since the support measures are locally determined and by Theorem 4.1.3 we obtain that

$$\begin{aligned}\mathbb{E}\left[\Phi_1^{0,2}\left(Z_{\alpha,\gamma,E},[0,1]^2\right)\right] &= \mathbb{E}\left[\Phi_1^{0,2}\left(Z_{\alpha,\gamma,E}\cap 2B^2,[0,1]^2\right)\right] = \overline{\Phi}_1^{0,2}(X)\,\mathrm{e}^{-\overline{V}_2(X)}\\ &= \overline{\Phi}_1^{0,2}(X)\,\mathrm{e}^{-\gamma V_2(E)}.\end{aligned}$$

Therefore, by (5.7), we get

$$\frac{\mathbb{E}\left[\left(\Phi_1^{0,2}\left(Z_{\alpha,\gamma,E},[0,1]^2\right)\right)_{1,1}\right]}{\mathbb{E}\left[\left(\Phi_1^{0,2}\left(Z_{\alpha,\gamma,E},[0,1]^2\right)\right)_{2,2}\right]} = \frac{\left(\overline{\Phi}_1^{0,2}(X)\right)_{1,1}}{\left(\overline{\Phi}_1^{0,2}(X)\right)_{2,2}} = \frac{(\alpha+1)\left(\Phi_1^{0,2}(E)\right)_{1,1} + \left(\Phi_1^{0,2}(E)\right)_{2,2}}{\left(\Phi_1^{0,2}(E)\right)_{1,1} + (\alpha+1)\left(\Phi_1^{0,2}(E)\right)_{2,2}}.$$

Hence, in the isotropic case ($\alpha = 0$) the ratio in (5.15) is equal to 1. Moreover, the quantity (5.15) is always independent of the volume fraction $\overline{V}_2(Z_{\alpha,\gamma,E})$, since the volume fraction depends by (5.11) only on the intensity γ and not on the parameter α. It is interesting and should be investigated further why these properties are also observed for the quantity $\langle\beta_1^{*\,0,2}\rangle$ in [STMK+11, Sect. 2.2].

Remark 5.4.3. For a smooth base grain $E \in C^2_+$ we obtain special formulas since the support measure Ξ_1 can be represented as an integral over the unit sphere weighted with the curvature radius of E (see (5.16)) or as an integral over the boundary of E (see (5.17)). We use the abbreviation $u(\alpha) := (\cos\alpha, \sin\alpha)^\top$, $\alpha \in \mathbb{R}$ and the notation $r(E,u)$ for the radius of curvature of E at a point $x \in \partial E$ with outer normal $u \in S^1$. The representations of the area respectively curvature measure for smooth convex bodies from Lemma 3.1.2, (ii) lead to

$$\Xi_1(E,\cdot) = \frac{1}{2}\int\limits_{S^1} \mathbf{1}\{(x(u),u) \in \cdot\} r(E,u)\mathcal{H}^1(du), \tag{5.16}$$

respectively

$$\Xi_1(E,\cdot) = \frac{1}{2}\int\limits_{\partial E} \mathbf{1}\{(x,u(x)) \in \cdot\}\mathcal{H}^1(dx), \tag{5.17}$$

where for $u \in S^1$ we denote by $x(u)$ the unique boundary point in ∂E with outer normal u and, for $x \in \partial E$, we denote by $u(x)$ the outer normal of E at x. If a parametrization of ∂E is known, (5.17) can be used to determine $\Phi_1^{0,s}(E)$ for $s \in \mathbb{N}_0$, and via equation (5.6) then also $\overline{\Phi}_1^{0,s}(X)$. On the other hand, (5.16) can be used to determine $V_{1,1}(\vartheta(\theta)E,E)$, and then via (5.8) also $\overline{V}_{1,1}(X,X)$; see (5.18). In fact, observe that it follows from (5.9) that

$$\begin{aligned}
V_{1,1}(\vartheta(\theta)E,E) &= \frac{1}{2\pi}\int\limits_0^{2\pi}\int\limits_0^{2\pi} \alpha\left(u(\beta_1-\beta_2+\theta),u(0)\right)|\sin(\beta_1-\beta_2+\theta)| \\
&\quad\times r(E,u(\beta_1))r(E,u(\beta_2))d\beta_1 d\beta_2 \\
&= \frac{1}{2\pi}\int\limits_0^{2\pi}\int\limits_0^{2\pi} \left(\mathbf{1}\{\beta_1 \in [0,\pi]\}\beta_1\sin(\beta_1) - \mathbf{1}\{\beta_1 \in (\pi,2\pi]\}(2\pi-\beta_1)\sin(\beta_1)\right) \\
&\quad\times r(E,u(\beta_1+\beta_2-\theta))r(E,u(\beta_2))d\beta_1 d\beta_2 \\
&= \frac{1}{2\pi}\int\limits_0^{2\pi}\int\limits_0^{\pi} \beta_1\sin(\beta_1)\left[r(E,u(\beta_1+\beta_2-\theta)) + r(E,u(-\beta_1+\beta_2-\theta))\right] \\
&\quad\times r(E,u(\beta_2))d\beta_1 d\beta_2,
\end{aligned}$$

and hence

$$\begin{aligned}
\overline{V}_{1,1}(X,X) &= \gamma^2 c(\alpha)^2 \int\limits_0^{2\pi}\int\limits_0^{2\pi} V_{1,1}(\vartheta(\theta_1)E,E)\,|\cos(\theta_1+\theta_2)|^\alpha|\cos(\theta_2)|^\alpha\, d\theta_1 d\theta_2 \\
&= \gamma^2 c(\alpha)^2 \int\limits_0^{2\pi}\int\limits_0^{2\pi} \frac{1}{2\pi}\int\limits_0^{2\pi}\int\limits_0^{\pi} \beta_1\sin(\beta_1)r(E,u(\beta_2))\left[r(E,u(\beta_1+\beta_2-\theta_1))\right.
\end{aligned}$$

$$+\, r(E,u(-\beta_1+\beta_2-\theta_1))\big]\,d\beta_1 d\beta_2\,|\cos(\theta_1+\theta_2)|^{\alpha}|\cos(\theta_2)|^{\alpha}\,d\theta_1 d\theta_2$$

$$= \frac{\gamma^2 c(\alpha)^2}{2\pi}\int_0^{2\pi}\int_0^{2\pi}\int_0^{2\pi}\int_0^{\pi} \beta_1\sin(\beta_1) r(E,u(\beta_2))\big[r(E,u(\beta_1+\beta_2-\theta_1+\theta_2))$$

$$+\, r(E,u(-\beta_1+\beta_2-\theta_1+\theta_2))\big]\,|\cos(\theta_1)|^{\alpha}|\cos(\theta_2)|^{\alpha} d\beta_1 d\beta_2\; d\theta_1 d\theta_2. \tag{5.18}$$

5.5 Planar Boolean model with smooth grains

In this subsection we consider a Boolean model Z with a grain distribution $\mathbb{Q}$ which is concentrated on $\mathcal{K}_0 \cap C^2_+$. We write again $u(\varphi) := (\cos\varphi, \sin\varphi)^\top$, $\varphi \in \mathbb{R}$. Then, we obtain from [Sch13a, (4.26)] and Fubini's theorem that

$$\overline{\Phi}_1^{0,s}(X) = c_{d-1}^{0,s}\gamma \int_{\mathcal{K}_0}\int_0^{2\pi} r(K,u(\varphi))u(\varphi)^s d\varphi\, \mathbb{Q}(dK)$$
$$= c_{d-1}^{0,s}\gamma \int_0^{2\pi}\int_{\mathcal{K}_0} r(K,u(\varphi))\mathbb{Q}(dK)u(\varphi)^s d\varphi,$$

where $r(K,u)$ is the radius of curvature of K at u, for $K \in \mathcal{K}_0 \cap C^2_+$ and $u \in S^1$, compare [Sch13a, (2.49)].

The surface tensor mean values are now related to the Fourier coefficients of the function $g : [0,2\pi] \to [0,\infty)$, where

$$g(\varphi) := \gamma \int_{\mathcal{K}_0} r(K,u(\varphi))\mathbb{Q}(dK).$$

We denote the sth Fourier coefficient of g by $\hat{g}(s)$. Then, we obtain for $s \in \mathbb{N}_0$ that

$$\hat{g}(s) = \frac{1}{2\pi}\int_0^{2\pi} g(\varphi)\mathrm{e}^{-\mathrm{i}s\varphi}d\varphi = \frac{1}{2\pi}\int_0^{2\pi} g(\varphi)(\cos(\varphi) - \mathrm{i}\sin(\varphi))^s d\varphi$$
$$= \sum_{j=0}^{s}\binom{s}{j}(-\mathrm{i})^{s-j}\frac{1}{2\pi}\int_0^{2\pi} g(\varphi)(\cos\varphi)^j(\sin\varphi)^{s-j}d\varphi$$
$$= \sum_{j=0}^{s}\binom{s}{j}(-\mathrm{i})^{s-j}\frac{s!\omega_{1+s}}{2\pi}\left(\overline{\Phi}_1^{0,s}(X)\right)_{\underbrace{1,\dots,1}_{j\text{ times}}\underbrace{2,\dots,2}_{s-j\text{ times}}}$$

and in the same way that

$$\hat{g}(-s) = \sum_{j=0}^{s} \binom{s}{j} \mathrm{i}^{s-j} \frac{s!\omega_{1+s}}{2\pi} \left(\overline{\Phi}_1^{0,s}(X)\right)_{\underbrace{1,\dots,1}_{j\text{ times}},\underbrace{2,\dots,2}_{s-j\text{ times}}}.$$

By the theorem of Carleson [Car66], it holds

$$\lim_{N\to\infty} \sum_{s=-N}^{s=N} \hat{g}(s)\mathrm{e}^{\mathrm{i}s\varphi} = g(\varphi)$$

for almost all $\varphi \in [0, 2\pi]$. Hence, it follows that the tensors

$$\overline{\Phi}_1^{0,s}(X), \quad s \in \mathbb{N}_0,$$

determine

$$\gamma\, \mathbb{E}[r(Z_0, u(\varphi))] \quad \text{for almost all } \varphi \in [0, 2\pi],$$

where Z_0 denotes the typical grain, i.e., a random convex body with distribution $\mathbb{Q}$.

Remark 5.5.1. The situation in higher dimensions is similar. Instead of just one radius of curvature one can use the product of all principal radii of curvature and the Fourier expansion can be replaced by an expansion into spherical harmonics.

5.6 Uniqueness results for Minkowski tensors

The coordinates of the translation invariant Minkowski tensors $\Phi_j^{0,s}$ are integrals of the area measure Ψ_j over spherical monomials of polynomial degree s. By the theorem of Stone-Weierstraß the spherical polynomials are a dense subset of $C(S^{n-1})$ with respect to the supremum norm. Thus, the measure $\overline{\Psi}_j(Z, \cdot)$ is uniquely determined by the sequence of densities of Minkowski tensors $\overline{\Phi}_j^{0,s}(Z)$, $s \in \mathbb{N}_0$. Hence, the inversion formulas for the densities of the area measures lead to uniqueness results for the densities of the Minkowski tensors.

In two dimensions we obtain the following uniqueness result.

Theorem 5.6.1. *Let Z be a stationary Boolean model in $\mathbb{R}^2$. Then, the densities*

$$\overline{V}_2(Z),\ \overline{\Phi}_1^{0,s}(Z), s \in \mathbb{N}_0 \text{ and } \overline{V}_0(Z)$$

determine uniquely

$$\mathbb{E}V_2(Z_0),\ \mathbb{E}\Phi_1^{0,s}(Z_0), s \in \mathbb{N}_0 \text{ and } \gamma.$$

Proof. By Corollary 4.2.4 we have

$$\gamma = \frac{1}{1-\overline{V}_2(Z)}\overline{V}_0(Z) + \frac{1}{2}\left(\frac{1}{1-\overline{V}_2(Z)}\right)^2 \int\limits_{S^1}\int\limits_{S^1} \mu_2(u_1,u_2;S^{d-1})\overline{\Psi}_1(Z,du_1)\overline{\Psi}_1(Z,du_2). \tag{5.19}$$

The measure $\overline{\Psi}_1(Z,\cdot)$ is uniquely determined by the sequence of densities of Minkowski tensors $\overline{\Phi}_1^{0,s}(Z)$, $s \in \mathbb{N}_0$. The assertion follows together with the well-known relations

$$\overline{V}_d(X) = -\ln(1-\overline{V}_d(Z)) \quad \text{and} \quad \overline{\Phi}_{d-1}^{0,s}(X) = \frac{1}{1-\overline{V}_d(Z)}\overline{\Phi}_{d-1}^{0,s}(Z). \tag{5.20}$$

□

In three dimensions we obtain a uniqueness result of the same nature.

Theorem 5.6.2. *Let Z be a stationary Boolean model in $\mathbb{R}^3$. Then, the densities*

$$\overline{V}_3(Z),\ \overline{\Phi}_2^{0,s}(Z),\ \overline{\Phi}_1^{0,s}(Z), s \in \mathbb{N}_0 \text{ and } \overline{V}_0(Z)$$

determine uniquely

$$\mathbb{E}V_3(Z_0),\ \mathbb{E}\Phi_2^{0,s}(Z_0),\ \mathbb{E}\Phi_1^{0,s}(Z_0), s \in \mathbb{N}_0 \text{ and } \gamma.$$

Proof. For $j \in \{1,2\}$ the measure $\overline{\Psi}_j(Z,\cdot)$ is uniquely determined by the densities of Minkowski tensors $\overline{\Phi}_j^{0,s}(Z)$, $s \in \mathbb{N}_0$. Hence, Corollary 4.2.5, Lemma 4.2.1 and (5.20) yield the assertion.

□

In dimensions higher than three it is an open problem wether the intensity can be determined from densities of Minkowski tensors of the Boolean model. Even if we include densities of mixed Minkowski tensors depending additionally on a unit vector u, the answer is not known. For example in the case $d = 4$ we have

$$\mathrm{mix}(0) = \{(0),(1,3),(3,1),(2,2),(2,3,3),(3,2,3),(3,3,2),(3,3,3)\}.$$

Thus, the density formula for $\overline{V}_0(Z)$ involves the mixed density $\overline{V}_{2,2}(X,X)$. This mixed density cannot be simplified using the decomposition results Corollary 3.4.3 or Lemma 3.4.4.

Though, if we include mixed Minkowski tensors depending on a unit vector we obtain the following result.

Theorem 5.6.3. *Let $d \geq 4$ and Z be a stationary Boolean model in $\mathbb{R}^d$. Then, the densities*

$$\overline{V}_d(Z), \overline{\Phi}^{0,s}_{d-1}(Z), \overline{\Phi}^{0,s}_{d-2}(Z), \overline{\Phi}^{0,s}_{d-3}(Z) \text{ and } \overline{\Phi}^{0,s}_{d-2}(Z,u), u \in S^{d-1}, s \in \mathbb{N}_0$$

determine uniquely

$$\overline{V}_d(X), \overline{\Phi}^{0,s}_{d-1}(X), \overline{\Phi}^{0,s}_{d-2}(X), \overline{\Phi}^{0,s}_{d-3}(X), s \in \mathbb{N}_0.$$

Proof. Since the measure $\overline{\Psi}_{d-1}(Z;\cdot)$ is uniquely determined by $\overline{\Phi}^{0,s}_{d-1}(Z), s \in \mathbb{N}_0$ the assertion follows from (5.20), Lemma 4.2.1 and Lemma 4.2.2. □

In comparison to the results in two and three dimensions Theorem 5.6.3 is unsatisfactory because densities of mixed Minkowski tensors depending on an arbitrarily chosen unit vector u are involved. Thus, uncountably many real-valued quantities have to be known. Furthermore, since the intensity cannot be determined, only a uniqueness result for densities of the particle process is obtained and not for mean values of the typical grain Z_0 as in two and three dimensions.

6 Application to harmonic intrinsic volumes

6.1 Harmonic intrinsic volumes

In this section we introduce a completely new collection of geometric functionals on $\mathcal{K}$, which we call *harmonic intrinsic volumes*. Each intrinsic volume is embedded into a sequence, where the first element is the ordinary intrinsic volume and the other elements are moments of the area measures with respect to orthonormal functions on the unit sphere. This definition is simple but leads in combination with the concept of rotation regularity to a surprising new perspective on the density formulas for the non-isotropic Boolean model.

It is a well-known result that in the isotropic situation the densities of the intrinsic volumes of the Boolean model can be expressed by the mean values of the intrinsic volumes of the typical grain and vice versa. These results are the Miles formulas, see (1.4) in two dimensions. So far in the non-isotropic situation extensions of the intrinsic volumes (like the mixed volumes) lead only to uniqueness results but not to explicit inversion results. In two and three dimensions the consideration of the harmonic intrinsic volumes will enable us to obtain expressions directly comparable to the results for intrinsic volumes in the isotropic situation. That is, we can express the densities of the harmonic intrinsic volumes of the Boolean model by the mean values of the harmonic intrinsic volumes of the typical grain and the other way round.

Moreover, if only a truncated sequence of the densities of harmonic intrinsic volumes of the Boolean model is used to approximate say the intensity, we will obtain error bounds in terms of the modulus of isotropy. The geometric functionals for which density formulas were derived previously are motivated by ideas from convex geometry. For instance the consideration of the volume of linear combinations of convex bodies leads to the mixed volumes 2.2 considered in [Wei01a], the investigation of characterization results for valuations on $\mathcal{K}$ with prescribed properties inspired the definition of the Minkowski tensors, which were considered in Chapter 5 and [HHKM14]. Also the consideration of densities of the centred support function in [Wei99] is inspired by geometric ideas. Our harmonic intrinsic volumes, on the other hand, are, apart from convex geometry, also inspired by harmonic analysis and approximation theory.

Recall from Section 2.4 that the functions $Y_{l,p}$, $1 \leq p \leq D(d,l)$, form an orthonormal basis of the space of spherical harmonics of polynomial degree $l \in \mathbb{N}_0$.

Definition 6.1.1. We define the *sequence of harmonic intrinsic volumes* of a polyconvex set $K \in \mathcal{R}$ by

$$V_j^{l,p}(K) := \int\limits_{S^{d-1}} Y_{l,p}(u)\Psi_j(K, du),$$

where $0 \leq j \leq d-1$, $l \in \mathbb{N}_0$ and $1 \leq p \leq D(d,l)$.

Observe that the harmonic intrinsic volumes depend on the choice of the basis B in Section 2.4.

For each $l \in \mathbb{N}_0$ we define the vector

$$V_j^l(K) := \left(V_j^{l,1}(K), \ldots, V_j^{l,D(d,l)}(K)\right).$$

The first element of the sequence of harmonic intrinsic volumes in lexographical order is

$$V_j^{0,1}(K) = V_j(K), \tag{6.1}$$

i.e. for $(l,p) = (0,1)$ the jth harmonic intrinsic volume is equal to the ordinary jth intrinsic volume. For all choices of (l,p) the harmonic intrinsic volume $V_j^{l,p}$ inherits various properties from the jth area measure, which we collect in the following lemma.

Lemma 6.1.2. *The harmonic intrinsic volume $V_j^{l,p}$ is positive homogeneous of degree j, additive, translation invariant and measurable on $\mathcal{R}$ and continuous on $\mathcal{K}$.*

In contrast to the intrinsic volumes the harmonic intrinsic volumes are not rotation invariant. However the following rotation formula holds.

Proposition 6.1.3. *Let $j \in \{0, \ldots, d-1\}$, $l \in \mathbb{N}_0$, $1 \leq p \leq D(d,l)$ and $K \in \mathcal{R}$. Then*

$$\int\limits_{SO_d} V_j^{l,p}(\vartheta K)\nu(d\vartheta) = \begin{cases} V_j(K), & (l,p) = (0,1), \\ 0, & \textit{otherwise}. \end{cases}$$

Proof. The relation for $(l,p) = (0,1)$ follows from (6.1) and the rotation invariance of the intrinsic volumes. For $(l,p) \neq (0,1)$ the rotation covariance of Ψ_j, which follows from Theorem 3.1.1, (vi) with $A = \mathbb{R}^d$, and Fubini's theorem imply

$$\begin{aligned} \int\limits_{SO_d} V_j^{l,p}(\vartheta K)\nu(d\vartheta) &= \int\limits_{SO_d}\int\limits_{S^{d-1}} Y_{l,p}(u)\Psi_j(\vartheta K, du)\nu(d\vartheta) \\ &= \int\limits_{S^{d-1}}\int\limits_{SO_d} Y_{l,p}(\vartheta u)\nu(d\vartheta)\Psi_j(K, du). \end{aligned}$$

Since σ is up to a constant the only rotation invariant measure on S^{d-1}, we get

$$\int_{SO_d} V_j^{l,p}(\vartheta K)\nu(d\vartheta) = \omega_d^{-1} \int_{S^{d-1}} \int_{S^{d-1}} Y_{l,p}(v)\sigma(dv)\Psi_j(K,du)$$
$$= \int_{S^{d-1}} (Y_{l,p}, Y_{0,1})\Psi_j(K,du) = 0.$$

□

For special convex bodies the harmonic intrinsic volumes have more explicit representations, which can be deduced from the corresponding representations of the area measures from Lemma 3.1.2. The harmonic intrinsic volume $V_j^{l,p}(P)$ of a polytope P is a weighted sum over the j-dimensional volumes of the j-faces of P. On the other hand, for a smooth convex body $K \in C_+^2$ the harmonic intrinsic volumes have an explicit representation as integrals of the jth normalized elementary symmetric function s_j of the principal radii of curvature of ∂K.

Lemma 6.1.4. *Let $j \in \{0, \dots, d-1\}$, $l \in \mathbb{N}_0$ and $1 \leq p \leq D(d,l)$.*

(i) *If P is a polytope, then*

$$V_j^{l,p}(P) = \frac{1}{\omega_{d-j}} \sum_{F \in \mathcal{F}_j(P)} \mathcal{H}^j(F) \int_{N(P,F)\cap S^{d-1}} Y_{l,p}(u)\mathcal{H}^{d-1-j}(du).$$

(ii) *If $K \in C_+^2$ is smooth, then*

$$V_j^{l,p}(K) = \frac{\binom{d-1}{j}}{\omega_{d-j}} \int_{S^{d-1}} Y_{l,p}(u)s_j(u)\sigma(du).$$

Furthermore, the situation for $j=0$ is special. Since the 0th area measure is proportional to the spherical Lebesgue measure, the orthonormality of the spherical harmonics yields

$$V_0^{l,p}(K) = \begin{cases} 1, & (l,p) = (0,1), \\ 0, & \text{otherwise}. \end{cases} \tag{6.2}$$

6.2 Mixed harmonic intrinsic volumes

We define mixed harmonic intrinsic volumes for $k \in \mathbb{N}, m \in \operatorname{mix}(j,k)$ and $K_1, \dots, K_k \in \mathcal{R}$ by

$$V_m^{l,p}(K_1, \dots, K_k) := \int_{S^{d-1}} Y_{l,p}(u)\Psi_m(K_1, \dots, K_k; du).$$

The translative integral formula for support measures Theorem 3.2.1 leads to the following translative integral formula for harmonic intrinsic volumes.

Corollary 6.2.1.

$$\int\limits_{(\mathbb{R}^d)^{k-1}} V_j^{l,p}(K_1 \cap (K_2+x_2) \cap \ldots \cap (K_k+x_k))\, d(x_2,\ldots,x_k) = \sum_{m\in \operatorname{mix}(j,k)} V_m^{l,p}(K_1,\ldots,K_k).$$

By Lemma 3.2.3 the mixed harmonic intrinsic volumes are measurable on $\mathcal{R}$. The mixed harmonic intrinsic volumes inherit various properties from the support measures, which we collect in a corollary.

Corollary 6.2.2. *Let $K_1,\ldots,K_k \in \mathcal{R}$ be polyconvex sets. Then*

(i) $V^{l,p}_{m_1,\ldots,m_k}(K_1,\ldots,K_k)$ *is symmetric with respect to permutations of* $\{1,\ldots,k\}$;

(ii) $V^{l,p}_{d,m_2,\ldots,m_k}(K_1,\ldots,K_k) = V_d(K_1)V^{l,p}_{m_2,\ldots,m_k}(K_2,\ldots,K_k)$;

(iii) $V^{l,p}_{m_1,\ldots,m_k}(K_1,\ldots,K_k)$ *is positively homogeneous of degree* m_i *with respect to* K_i *for* $i \in \{1,\ldots,k\}$;

(iv) *if* $K_1,\ldots,K_k$ *are polytopes, then*

$$\begin{aligned} &V^{l,p}_{m_1,\ldots,m_k}(K_1,\ldots,K_k)\\ &\quad= \frac{1}{\omega_{d-j}} \sum_{F_1\in\mathcal{F}_{m_1}(K_1)} \cdots \sum_{F_k\in\mathcal{F}_{m_k}(K_k)} \int\limits_{\left(\sum_{i=1}^k N(K_i,F_i)\right)\cap S^{d-1}} Y_{l,p}(u)\,\mathcal{H}^{d-1-j}(du)\\ &\qquad\times [F_1,\ldots,F_k]\,\mathcal{H}^{m_1}(F_1)\,\mathcal{H}^{m_2}(F_2)\cdots\mathcal{H}^{m_k}(F_k); \end{aligned}$$

(v) *the map* $V_m^{l,p}$ *is additive and continuous on* $\mathcal{K}^k$ *and measurable on* $\mathcal{R}^k$;

(vi)

$$V_m^{l,p}(K_1+x_1,\ldots,K_k+x_k) = V_m^{l,p}(K_1,\ldots,K_k), \quad x_1,\ldots,x_k \in \mathbb{R}^d;$$

(vii)

$$V_m^{l,p}(\sigma K_1,\ldots,\sigma K_k) = V_m^{l,p}(K_1,\ldots,K_k), \quad \sigma \in SO_d.$$

6.3 Mean value and density formulas for harmonic intrinsic volumes

In this section we first establish connections between mean values of the harmonic intrinsic volumes of the intersection of Z with a compact, convex window W and the densities of the particle process X. In a second step we obtain corresponding relations between the densities of harmonic intrinsic volumes of the Boolean model and the densities of harmonic intrinsic volumes of the underlying particle process. As the special case $(l,p) = (0,1)$ one obtains the well-known formulas for the densities of intrinsic volumes of a non-isotropic Boolean model.

6.3.1 The isotropic situation

Under the assumption of isotropy the mean harmonic intrinsic volume of a Boolean model intersected with a ball is equal to zero in all cases except for $(l,p) = (0,1)$.

Proposition 6.3.1. *Let Z be an isotropic Boolean model, $j \in \{0,\dots,d-1\}$, $l \in \mathbb{N}_0$ and $1 \le p \le D(d,l)$ and $r > 0$. Then*

$$\mathbb{E}\, V_j^{l,p}\left(Z \cap rB^d\right) = 0, \qquad (l,p) \neq (0,1)$$

and

$$\overline{V}_j^{l,p}(Z) = \begin{cases} \overline{V}_j(Z), & (l,p) = (0,1) \\ 0, & \textit{otherwise}. \end{cases}$$

Proof. For $(l,p) \neq (0,1)$ the isotropy of Z, Fubini's theorem and Proposition 6.1.3 imply

$$\begin{aligned} \mathbb{E} V_j^{l,p}(Z \cap rB^d) &= \int_{SO_d} \mathbb{E}\left[V_j^{l,p}(\vartheta Z \cap rB^d)\right] \nu(d\vartheta) \\ &= \mathbb{E} \int_{SO_d} V_j^{l,p}\left(\vartheta(Z \cap rB^d)\right) \nu(d\vartheta) = 0. \end{aligned}$$

The relation for $\overline{V}_j^{l,p}(Z)$ is an immediate consequence. □

6.3.2 The non-isotropic situation

For a non-isotropic Boolean model we obtain the following mean value formulas for harmonic intrinsic volumes as a corollary of Theorem 4.1.3.

Theorem 6.3.2. *Let Z be a stationary Boolean model, $W \in \mathcal{K}$, $0 \leq j \leq d-1$, $l \in \mathbb{N}_0$ and $1 \leq p \leq D(d,l)$. Then*

$$\mathbb{E}\left[V_j^{l,p}(Z \cap W)\right] = V_j^{l,p}(W) - \mathrm{e}^{-\overline{V}_d(X)} \sum_{m \in \mathrm{mix}(j)} \frac{(-1)^{|m|-1}}{|m|!}\Big[|m|\overline{V}_m^{l,p}(W, X, \ldots, X) - V_d(W)\overline{V}_m^{l,p}(X, \ldots, X)\Big].$$

As a consequence of the previous result we obtain density formulas for the harmonic intrinsic volumes in general dimension.

Corollary 6.3.3. *Let Z be a stationary Boolean model, $0 \leq j \leq d-1$, $l \in \mathbb{N}_0$ and $1 \leq p \leq D(d,l)$. Then*

$$\overline{V}_j^{l,p}(Z) = \mathrm{e}^{-\overline{V}_d(X)} \sum_{m \in \mathrm{mix}(j)} \frac{(-1)^{|m|-1}}{|m|!} \overline{V}_m^{l,p}(X, \ldots, X).$$

6.3.3 Density formulas for mixed harmonic intrinsic volumes

We define mixed harmonic intrinsic volumes depending on unit vectors, which are derived from the special mixed area measures of Definition 3.2.4.

Let $k \in \mathbb{N}$, $j \in \{0, \ldots, d-1\}$, $1 \leq k' \leq d-j$, $u_1, \ldots, u_{k'} \in S^{d-1}$, $K_1, \ldots, K_k \in \mathcal{K}$, $m \in \mathrm{mix}(k'+j, k)$, $l \in \mathbb{N}_0$ and $1 \leq p \leq D(d,l)$. Then we define

$$V_m^{l,p}(K_1, \ldots, K_k, u_1, \ldots, u_{k'}) := 2^{k'} \int_{S^{d-1}} Y_{l,p}(v)\Psi_m\Big(K_1, \ldots, K_k, u_1, \ldots, u_{k'}; dv\Big).$$

Then, the following density formulas for the mixed harmonic intrinsic volumes follow directly from Theorem 4.1.5.

Corollary 6.3.4. *Let Z be a stationary Boolean model, $L_1, \ldots, L_k \in \mathcal{R}$, $j \in \{0, \ldots, d-1\}$, $j \leq j' \leq d-1$, $1 \leq k \leq d-j$, $m \in \mathrm{mix}(d-j'+j, k)$, $l \in \mathbb{N}_0$ and $1 \leq p \leq D(d,l)$. Then,*

$$\overline{V}_{j',m}^{l,p}(Z, L_1, \ldots, L_k) = \mathrm{e}^{-\overline{V}_d(X)} \sum_{m' \in \mathrm{mix}(j')} \frac{(-1)^{|m'|-1}}{|m'|!} \overline{V}_{m',m}^{l,p}(X, \ldots, X, L_1, \ldots, L_k) \quad (6.3)$$

and for $u_1, \ldots, u_k \in S^{d-1}$ we get

$$\overline{V}_{k+j,m}^{l,p}(Z, u_1, \ldots, u_k) = \mathrm{e}^{-\overline{V}_d(X)} \sum_{m' \in \mathrm{mix}(k+j)} \frac{(-1)^{|m'|-1}}{|m'|!} \overline{V}_{m',m}^{l,p}(X, \ldots, X, u_1, \ldots, u_k). \quad (6.4)$$

6.4 Uniqueness results for harmonic intrinsic volumes

The harmonic intrinsic volume $V_j^{l,p}$ is the integral over the spherical harmonic $Y_{l,p}$ with respect to the area measure Ψ_j. The spherical harmonics are a dense subset of $L^2(S^{d-1})$. Thus, the measure $\overline{\Psi}_j(Z, \cdot)$ is uniquely determined by the sequence of densities of harmonic intrinsic volumes

$$\overline{V}_j^{l,p}(Z) = \int\limits_{S^{d-1}} Y_j^{l,p}(u)\overline{\Psi}_j(Z, du), \quad l \in \mathbb{N}_0, 1 \leq p \leq D(d,l).$$

Hence, we can obtain uniqueness results based on the inversion formulas for the densities of area measures. The proofs are exactly as for the Minkowski tensors, which is why we omit them. In two dimensions we obtain the following statement.

Theorem 6.4.1. *Let Z be a stationary Boolean model in $\mathbb{R}^2$. Then, the densities*

$$\overline{V}_2(Z),\ \overline{V}_1(Z),\ \overline{V}_1^{l,p}(Z), l \in \mathbb{N}, 1 \leq p \leq 2 \text{ and } \overline{V}_0(Z)$$

determine uniquely

$$\mathbb{E}V_2(Z_0),\ \mathbb{E}V_1(Z_0),\ \mathbb{E}V_1^{l,p}(Z_0), l \in \mathbb{N}, 1 \leq p \leq 2 \text{ and } \gamma.$$

Also in three dimensions the mean values of the harmonic intrinsic volumes of the typical grain are determined by the corresponding densities of the Boolean model.

Theorem 6.4.2. *Let Z be a stationary Boolean model in $\mathbb{R}^3$. Then, the densities*

$$\overline{V}_3(Z),\ \overline{V}_2^{l,p}(Z),\ \overline{V}_1^{l,p}(Z), l \in \mathbb{N}_0, 1 \leq p \leq 2l+1 \text{ and } \overline{V}_0(Z)$$

determine uniquely

$$\mathbb{E}V_3(Z_0),\ \mathbb{E}V_2^{l,p}(Z_0),\ \mathbb{E}V_1^{l,p}(Z_0), l \in \mathbb{N}_0, 1 \leq p \leq 2l+1 \text{ and } \gamma.$$

In dimensions higher than three we meet the same problem as for the Minkowski tensors. That is, we cannot determine the mean values of the harmonic intrinsic volumes of Z_0 from densities of harmonic intrinsic volumes of the Boolean model. Even if we include densities of mixed harmonic intrinsic volumes depending additionally on unit vectors, the answer is not known. For example in the case $n = 4$ we have

$$\text{mix}(0) = \{(0), (1,3), (3,1), (2,2), (2,3,3), (3,2,3), (3,3,2), (3,3,3)\}.$$

Thus, the density formula for $\overline{V}_0(Z)$ involves the mixed density $\overline{V}_{2,2}(X, X)$, for which a representation separating the two arguments in a way comparable to Corollary 3.4.3 or Lemma 3.4.4 is missing. However, we can at least formulate the following uniqueness result for the densities of harmonic intrinsic volumes.

Theorem 6.4.3. *Let $d \geq 4$ and Z be a stationary Boolean model. Then, the densities*

$$\overline{V}_d(Z), \overline{V}^{l,p}_{d-1}(Z), \overline{V}^{l,p}_{d-2}(Z), \overline{V}^{l,p}_{d-3}(Z) \text{ and } \overline{V}^{l,p}_{d-2}(Z,u),$$

with $u \in S^{d-1}, l \in \mathbb{N}_0, 1 \leq p \leq D(d,l)$ determine uniquely

$$\overline{V}_d(X), \overline{V}^{l,p}_{d-1}(X), \overline{V}^{l,p}_{d-2}(X), \overline{V}^{l,p}_{d-3}(X), l \in \mathbb{N}_0, 1 \leq p \leq D(d,l).$$

7 The method of densities via harmonic intrinsic volumes

For an isotropic, stationary Boolean model the density formulas for intrinsic volumes can be inverted using the classical method of moments (see e.g. [Mol97]).

In the isotropic and stationary case the densities of the Minkowski tensors are by Proposition 5.3.1 proportional to the densities of the intrinsic volumes and thus they do not contain additional information. On the other hand, the densities of the harmonic intrinsic volumes are all equal to zero in the isotropic situation except for those which are exactly the densities of the intrinsic volumes. For a non-isotropic stationary Boolean model an inversion of the density formulas for Minkowski tensors, which were obtained in Corollary 5.3.3, is in general not possible because mixed terms occur. However, in two and three dimensions uniqueness results in Theorem 5.6.1 and Theorem 5.6.2 hold, which show that the sequence of all densities of the Minkowski tensors determines the mean Minkowski tensors of the typical grain. Results of the same type have been obtained for the harmonic intrinsic volumes. At first sight these uniqueness results seem quite appealing but actually they are purely theoretical. In fact, it is not clear how they can be used to directly determine the intensity, the mean Minkowski tensors, or the mean harmonic intrinsic volumes of the typical grain. For this purpose it is necessary to obtain explicit inversion formulas, i.e. formulas expressing say the intensity in terms of countably many densities of geometric functionals. Results of this kind can be obtained if we make a minor regularity assumption on the intensity measure. Namely, we consider a stationary setting which is located between isotropy and anisotropy. We introduce a *rotation-regular* particle process (see Definition 7.2.2), that is a particle process for which the grain distribution fulfils a disintegration property with respect to the Haar measure on SO_d. A density $\eta : \mathcal{K}_0 \times SO_d \to [0, \infty)$ occurs on which smoothness assumptions can be made. It is then possible to introduce a modulus of isotropy expressing how large the deviation from isotropy is. For a Boolean model obtained from a rotation regular particle process the density formulas for the harmonic intrinsic volumes can be completely inverted. That is the densities of the harmonic intrinsic volumes of the particle process can be expressed in terms of the densities of the harmonic intrinsic volumes of the Boolean model (see Theorem 7.7.1 for the result in two dimensions and Theorem 7.7.6 for the result in three dimensions). Still, in practise these inversion results cannot be applied directly, since infinite sums of densities of the harmonic intrinsic volumes are involved. We derive error bounds if only finitely many densities of harmonic intrinsic volumes are known. The error bounds depend on the previously introduced modulus of isotropy.

7.1 A disintegration result

The following theorem provides a unique disintegration of every probability measure on the space of centered convex bodies with respect to a suitable rotation invariant measure. The disintegration is made unique by imposing a shift property on the occurring probability kernel.

Theorem 7.1.1. *Let $\mathbb{Q}$ be a probability measure on $\mathcal{K}_0$. Then, there is a rotation invariant probability measure $\tilde{\mathbb{Q}}$ on $\mathcal{K}_0$ and a probability kernel $\rho : \mathcal{K}_0 \times \mathcal{B}(SO_d) \to [0, \infty)$ which has for all $K \in \mathcal{K}_0$ and $\vartheta \in SO_d$ the property*

$$\rho(\vartheta K, B) = \rho(K, B\vartheta), \quad B \in \mathcal{B}(SO_d) \tag{7.1}$$

and fulfils

$$\mathbb{Q}(A) = \int_{\mathcal{K}_0} \int_{SO_d} \mathbf{1}\{\vartheta K \in A\} \rho(K, d\vartheta) \tilde{\mathbb{Q}}(dK), \quad A \in \mathcal{B}(\mathcal{K}_0). \tag{7.2}$$

The probability measure $\tilde{\mathbb{Q}}$ is unique and the probability kernel ρ is unique $\tilde{\mathbb{Q}}$–almost everywhere provided that (7.1) *and* (7.2) *are satisfied.*

Proof. The main idea of the proof is to find a measure μ on $\mathcal{K}_0 \times SO_d$ such that $\mathbb{Q}$ is the image measure of μ under the mapping $(K, \vartheta) \mapsto \vartheta K$ and $\tilde{\mathbb{Q}}$ is the marginal measure of μ on $\mathcal{K}_0$. A disintegration of μ implies then the relation (7.2) with a probability kernel ρ.

We define a measure $\tilde{\mathbb{Q}}$ on $\mathcal{K}_0$ by

$$\tilde{\mathbb{Q}} := \int_{\mathcal{K}_0} \int_{SO_d} \mathbf{1}\{\vartheta K \in \cdot\} \nu(d\vartheta) \mathbb{Q}(dK). \tag{7.3}$$

The rotation invariance of $\tilde{\mathbb{Q}}$ follows from the rotation invariance of ν. Now, define a measure μ on $\mathcal{K}_0 \times SO_d$ by

$$\mu := \int_{\mathcal{K}_0} \int_{SO_d} \mathbf{1}\{(\vartheta K, \vartheta^{-1}) \in \cdot\} \nu(d\vartheta) \mathbb{Q}(dK). \tag{7.4}$$

Obviously, μ and $\tilde{\mathbb{Q}}$ are related by

$$\tilde{\mathbb{Q}} = \mu(\cdot \times SO_d). \tag{7.5}$$

Furthermore, we have

$$\mathbb{Q} = \int_{\mathcal{K}_0} \int_{SO_d} \mathbf{1}\{\vartheta^{-1}(\vartheta K) \in \cdot\}\nu(d\vartheta)\mathbb{Q}(dK)$$
$$= \int_{\mathcal{K}_0 \times SO_d} \mathbf{1}\{\vartheta K \in \cdot\}\mu(d(K,\vartheta)). \tag{7.6}$$

Now, the theorem about the existence of the conditional distribution ([Kal97, Theorem 5.3]) implies the existence of a $\tilde{\mathbb{Q}}$-almost everywhere unique probability kernel ρ' from $\mathcal{K}_0$ to SO_d with

$$\mu(A \times B) = \int_A \rho'(K,B)\tilde{\mathbb{Q}}(dK), \quad A \in \mathcal{B}(\mathcal{K}_0), B \in \mathcal{B}(SO_d). \tag{7.7}$$

The probability kernel ρ' fulfils (7.2) but not necessarily (7.1). To obtain a probability kernel which fulfils also (7.1) we construct a smoothed version of ρ'. Namely, we define a probability kernel ρ from $\mathcal{K}_0$ to SO_d by

$$\rho(K,B) := \int_{SO_d} \rho'(\varrho K, B\varrho^{-1})\nu(d\varrho), \quad B \in \mathcal{B}(SO_d), K \in \mathcal{K}_0.$$

The definition (7.4) of μ and the rotation invariance of ν imply

$$\mu(\varrho A \times B\varrho^{-1}) = \mu(A \times B), \quad \text{for all } \varrho \in SO_d.$$

This, the rotation invariance of $\tilde{\mathbb{Q}}$ and Fubini's theorem imply

$$\mu(A \times B) = \int_{SO_d} \mu(\varrho A \times B\varrho^{-1})\nu(d\varrho) = \int_{SO_d} \int_A \rho'(\varrho K, B\varrho^{-1})\tilde{\mathbb{Q}}(dK)\nu(d\varrho)$$
$$= \int_A \rho(K,B)\tilde{\mathbb{Q}}(dK).$$

Moreover, the kernel ρ fulfils the property (7.1) since the rotation invariance of ν yields for every $\vartheta \in SO_d$, $B \in \mathcal{B}(SO_d)$ and $K \in \mathcal{K}_0$ that

$$\rho(\vartheta K, B) = \int_{SO_d} \rho'\left(\varrho\vartheta K, B\varrho^{-1}\right)\nu(d\varrho) = \int_{SO_d} \rho'\left(\varrho K, B\vartheta\varrho^{-1}\right)\nu(d\varrho) = \rho(K, B\vartheta).$$

In order to show the uniqueness we assume for $i \in \{1,2\}$ that $\tilde{\mathbb{Q}}_i$ is a rotation invariant probability measure on $\mathcal{K}_0$ and ρ_i a probability kernel from $\mathcal{K}_0$ to SO_d

such that (7.1) and (7.2) are fulfilled. For $i \in \{1,2\}$ we define a measure μ_i on $\mathcal{K}_0 \times SO_d$ by

$$\mu_i(A \times B) := \int_A \rho_i(K,B)\tilde{\mathbb{Q}}_i(dK), \quad A \in \mathcal{B}(\mathcal{K}_0), B \in \mathcal{B}(SO_d).$$

The mapping $T : (K,\vartheta) \mapsto (\vartheta K, K^{-1})$ on $\mathcal{K}_0 \times SO_d$ is measurable and bijective. For $i \in \{1,2\}$ the rotation invariance of $\tilde{\mathbb{Q}}_i$, the property (7.1), Fubini's theorem and the relation (7.2) imply for $A \in \mathcal{B}(\mathcal{K}_0), B \in \mathcal{B}(SO_d)$ that

$$\begin{aligned}
\mu_i(T(A \times B)) &= \int_{\mathcal{K}_0}\int_{SO_d} \mathbf{1}\{(\vartheta K, \vartheta^{-1}) \in A \times B\}\rho_i(K,d\vartheta)\tilde{\mathbb{Q}}_i(dK) \\
&= \int_{SO_d}\int_{\mathcal{K}_0}\int_{SO_d} \mathbf{1}\{(\vartheta \sigma K, \vartheta^{-1}) \in A \times B\}\rho_i(\sigma K,d\vartheta)\tilde{\mathbb{Q}}_i(dK)\nu(d\sigma) \\
&= \int_{SO_d}\int_{\mathcal{K}_0}\int_{SO_d} \mathbf{1}\{(\vartheta K, \sigma\vartheta^{-1}) \in A \times B\}\rho_i(K,d\vartheta)\tilde{\mathbb{Q}}_i(dK)\nu(d\sigma) \\
&= \int_{SO_d}\int_{\mathcal{K}_0}\int_{SO_d} \mathbf{1}\{\vartheta K \in A\} \int_{SO_d} \mathbf{1}\{\sigma\vartheta^{-1} \in B\}\nu(d\sigma)\rho_i(K,d\vartheta)\tilde{\mathbb{Q}}_i(dK) \\
&= \int_{\mathcal{K}_0}\int_{SO_d} \mathbf{1}\{\vartheta K \in A\}\rho_i(K,d\vartheta)\tilde{\mathbb{Q}}_i(dK)\nu(B) \\
&= \mathbb{Q}(A)\nu(B).
\end{aligned}$$

Thus, $\mu_1 \circ T = \mu_2 \circ T$, which yields $\mu_1 = \mu_2$. This implies $\tilde{\mathbb{Q}}_1 = \tilde{\mathbb{Q}}_2$ and $\tilde{\mathbb{Q}}_1$-almost everywhere $\rho_1 = \rho_2$. □

Remark 7.1.2. In the statement of Theorem 7.1.1 the measure $\mathbb{Q}$ lives on the space of convex bodies. This assumption can be relaxed considerably. The proof carries over to the more general setting of a compact group G with countable basis operating continuously on a Hausdorff-space E with countable basis. Thus, a probability measure $\mathbb{Q}$ on E has a unique disintegration into a G-invariant probability measure $\tilde{\mathbb{Q}}$ on E and a kernel ρ fulfilling an invariance property of the form (7.1). For example a measure $\mathbb{Q}$ on the space of centered compact sets $\mathcal{C}_0$ fulfils a corresponding disintegration result.

Remark 7.1.3. For a probability measure $\mathbb{Q}$ on $\mathcal{K}_0$ we call the unique measure $\tilde{\mathbb{Q}}$ from Theorem 7.1.1 the rotation invariant part of $\mathbb{Q}$. By (7.3) the rotation invariant part of $\mathbb{Q}$ has the explicit representation

$$\tilde{\mathbb{Q}}(A) = \int_{\mathcal{K}_0}\int_{SO_d} \mathbf{1}\{\vartheta K \in A\}\nu(d\vartheta)\mathbb{Q}(dK), \quad A \in \mathcal{B}(\mathcal{K}_0). \tag{7.8}$$

Remark 7.1.4. The disintegration of measures which are invariant under the action of a group acting on the measure space is an active field of research, see [Kal11, GL11] and the references therein. If we replace the probability kernel ρ by the probability kernel

$$\tilde{\rho}(K, B) := \rho(K, B^{-1}), \quad K \in \mathcal{K}_0, B \in \mathcal{B}(SO_d),$$

we obtain the following equivalent formulation of Theorem 7.1.1.

Let $\mathbb{Q}$ be a probability measure on $\mathcal{K}_0$ and $\tilde{\mathbb{Q}}$ its rotation invariant part. Then, there is a $\tilde{\mathbb{Q}}$-almost everywhere unique probability kernel $\tilde{\rho}$ from $\mathcal{K}_0$ to SO_d which is jointly rotation invariant, i.e.

$$\tilde{\rho}(\vartheta K, \vartheta B) = \tilde{\rho}(K, B), \quad B \in \mathcal{B}(SO_d), K \in \mathcal{K}_0, \vartheta \in SO_d$$

and satisfies

$$\mathbb{Q} = \int_{\mathcal{K}_0} \int_{SO_d} \mathbf{1}\{\vartheta^{-1} K \in \cdot\} \tilde{\rho}(K, d\vartheta) \tilde{\mathbb{Q}}(dK).$$

The above statement can be embedded in the general theory of invariant measures. Namely, we define by

$$\tilde{\mu} := \int_{\mathcal{K}_0} \int_{SO_d} \mathbf{1}\{(\vartheta K, \vartheta) \in \cdot\} \nu(d\vartheta) \mathbb{Q}(dK)$$

a rotation invariant measure on $\mathcal{K}_0 \times SO_d$, which satisfies

$$\mathbb{Q} = \int_{\mathcal{K}_0 \times SO_d} \mathbf{1}\{\vartheta^{-1} K \in \cdot\} \tilde{\mu}(d(K, \vartheta)).$$

An application of the general disintegration result [Kal07, Theorem 3.5] to the rotation invariant measure $\tilde{\mu}$ implies now the existence of the rotation invariant probability kernel $\tilde{\rho}$.

Remark 7.1.5. Combining results from the general theory of invariant measures one obtains even an explicit representation of the probability kernel ρ.

Firstly, Kallenberg showed in [Kal11, Theorem 2.4] that we can select representators of the orbits

$$SO_d K := \{\vartheta K : \vartheta \in SO_d\}, \quad K \in \mathcal{K}_0$$

in a u-measurable way, i.e. there is a measurable mapping $\alpha : \mathcal{K}_0 \to \mathcal{K}_0$ which has for $\tilde{\mathbb{Q}}$-almost all $K \in \mathcal{K}_0$ the properties

$$\alpha(K) \in SO_d K \text{ and } \alpha(K) = \alpha(L) \text{ for } \tilde{\mathbb{Q}}\text{-almost all } L \in SO_d K.$$

Let

$$\hat{\mathbb{Q}} := \int\limits_{\mathcal{K}_0} \mathbf{1}\{\alpha(K) \in \cdot\}\mathbb{Q}(dK).$$

Then, [Kal97, Theorem 5.3] implies the existence of a probability kernel q from $\mathcal{K}_0$ to SO_d such that

$$\mu = \int\limits_{\mathcal{K}_0} \int\limits_{SO_d} \mathbf{1}\{(K, \vartheta) \in \cdot\}q(K, d\vartheta)\hat{\mathbb{Q}}(dK).$$

Secondly, we need the so-called inversion kernel introduced for example in Kallenberg [Kal11, Theorem 3.1] or Gentner and Last [GL11, Theorem 2.1]. For $K, L \in \mathcal{K}_0$ let

$$SO_d(K, L) := \{\vartheta \in SO_d : \vartheta K = L\}.$$

Then, the inversion kernel is the unique probability kernel i from $\mathcal{K}_0$ to SO_d which has for $K \in \mathcal{K}_0$ the properties

$$i(\vartheta K, \vartheta B) = i(K, B), \quad B \in \mathcal{B}(SO_d), \vartheta \in SO_d$$

and

$$i(K, SO_d(\alpha(K), K)) = 1.$$

Now, we can show by the uniqueness of the kernel ρ from Theorem 7.1.1 that ρ coincides $\tilde{\mathbb{Q}}$-almost everywhere with

$$\rho'(K, \cdot) := \int\limits_{SO_d} \int\limits_{SO_d} \mathbf{1}\{\vartheta\sigma^{-1} \in \cdot\}q(\alpha(K), d\vartheta)i(K, d\sigma), \quad K \in \mathcal{K}_0.$$

Namely, ρ' fulfils (7.1) since the rotation invariance of i and α implies for $\vartheta \in SO_d$ and $B \in \mathcal{B}(\mathcal{K}_0)$ that

$$\begin{aligned}\rho'(\vartheta K, B\vartheta^{-1}) &= \int\limits_{SO_d} q(\alpha(\vartheta K), B\vartheta^{-1}\sigma)i(\vartheta K, d\sigma)\\ &= \int\limits_{SO_d} q(\alpha(K), B\vartheta^{-1}\vartheta\sigma)i(K, d\sigma) = \rho'(K, B).\end{aligned}$$

Furthermore, ρ' fulfils (7.2) since $i(K, \cdot)$ is concentrated on $SO_d(\alpha(K), K)$ and since the definition of $\tilde{\mathbb{Q}}$, the rotation invariance of α and the definition of $\hat{\mathbb{Q}}$ and q imply for $A \in \mathcal{B}(\mathcal{K}_0)$ that

$$\begin{aligned}&\int\limits_{\mathcal{K}_0} \int\limits_{SO_d} \mathbf{1}\{\vartheta K \in A\}\rho'(K, d\vartheta)\tilde{\mathbb{Q}}(dK)\\ &= \int\limits_{\mathcal{K}_0} \int\limits_{SO_d} \int\limits_{SO_d} \mathbf{1}\{\vartheta\sigma^{-1}K \in A\}q(\alpha(K), d\vartheta)i(K, d\sigma)\tilde{\mathbb{Q}}(dK)\end{aligned}$$

$$
\begin{aligned}
&= \int\limits_{\mathcal{K}_0} \int\limits_{SO_d} \mathbf{1}\{\vartheta\alpha(K) \in A\} q(\alpha(K), d\vartheta) \tilde{\mathbb{Q}}(dK) \\
&= \int\limits_{SO_d} \int\limits_{\mathcal{K}_0} \int\limits_{SO_d} \mathbf{1}\{\vartheta\alpha(\varrho K) \in A\} q(\alpha(\varrho K), d\vartheta) \mathbb{Q}(dK) \nu(d\varrho) \\
&= \int\limits_{\mathcal{K}_0} \int\limits_{SO_d} \mathbf{1}\{\vartheta\alpha(K) \in A\} q(\alpha(K), d\vartheta) \mathbb{Q}(dK) \\
&= \int\limits_{\mathcal{K}_0} \int\limits_{SO_d} \mathbf{1}\{\vartheta K \in A\} q(K, d\vartheta) \hat{\mathbb{Q}}(dK) \\
&= \mathbb{Q}(A).
\end{aligned}
$$

Now, it follows from the uniqueness of the kernel ρ from Theorem 7.1.1 for $\tilde{\mathbb{Q}}$-almost all $K \in \mathcal{K}_0$ that

$$
\rho(K, \cdot) = \int\limits_{SO_d} \int\limits_{SO_d} \mathbf{1}\{\vartheta\sigma^{-1} \in \cdot\} q(\alpha(K), d\vartheta) i(K, d\sigma). \tag{7.9}
$$

Alternatively, (7.9) can be derived via the technique of skew factorization described in [Kal11]. For this purpose, one has to apply [Kal11, Theorem 4.4] to the invariant measure $\tilde{\mu}$ introduced in Remark 7.1.4.

Example 7.1.6. If already the measure $\mathbb{Q}$ is rotation invariant, the uniqueness in the statement of Theorem 7.1.1 implies $\tilde{\mathbb{Q}} = \mathbb{Q}$ and

$$
\rho(K, \cdot) = \nu, \quad K \in \mathcal{K}_0.
$$

Example 7.1.7. Let $m \in \mathbb{N}$, $E_1, \dots, E_m \in \mathcal{K}_0$ with $SO_d E_i \neq SO_d E_j$ for $i \neq j$, $a_1, \dots, a_m > 0$ with $\sum_{i=1}^m a_i = 1$, $q_1, \dots, q_m$ probability measures on SO_d and

$$
\mathbb{Q} = \sum_{i=1}^m a_i \int\limits_{SO_d} \mathbf{1}\{\vartheta E_i \in \cdot\} q_i(d\vartheta).
$$

Then, it holds

$$
\tilde{\mathbb{Q}} = \sum_{i=1}^m a_i \int\limits_{SO_d} \mathbf{1}\{\vartheta E_i \in \cdot\} \nu(d\vartheta)
$$

and the kernel ρ fulfilling the properties of Theorem 7.1.1 is by Remark 7.1.5 for $\tilde{\mathbb{Q}}$-almost all $K \in \mathcal{K}_0$ and Borel sets $B \subset SO_d$ given by

$$
\rho(K, B) = \int\limits_{SO_d} q_i(B\sigma) i(K, d\sigma), \quad K \in SO_d E_i, 1 \le i \le m.
$$

7.2 Rotation regularity

A probability measure $\mathbb{Q}$ on $\mathcal{K}_0$ is called rotation regular if it has a decomposition of the form

$$\mathbb{Q}(A) = \int_{\mathcal{K}_0} \int_{SO_d} \mathbf{1}\{\vartheta K \in A\} \eta(K, \vartheta) \nu(d\vartheta) \tilde{\mathbb{Q}}(dK), \quad A \in \mathcal{B}(\mathcal{K}_0), \tag{7.10}$$

with a rotation invariant probability measure $\tilde{\mathbb{Q}}$ on $\mathcal{K}_0$ and a measurable function $\eta \geq 0$ on $\mathcal{K}_0 \times SO_d$ satisfying

$$\int_{SO_d} \eta(K, \vartheta) \nu(d\vartheta) = 1, \quad K \in \mathcal{K}_0 \tag{7.11}$$

and

$$\eta(\varrho K, \vartheta) = \eta(K, \vartheta \varrho), \quad K \in \mathcal{K}_0, \vartheta, \varrho \in SO_d. \tag{7.12}$$

The following theorem collects several statements which are equivalent to rotation regularity. Especially Theorem 7.2.1, (a) shows that rotation regularity is fulfilled for a large class of probability measures on $\mathcal{K}_0$.

Theorem 7.2.1. *Let $\mathbb{Q}$ be a probability measure on $\mathcal{K}_0$. Then, $\mathbb{Q}$ has a disintegration as in Theorem 7.1.1 with a rotation invariant measure $\tilde{\mathbb{Q}}$ and a kernel ρ from $\mathcal{K}_0$ to SO_d satisfying (7.1). The rotation regularity of $\mathbb{Q}$ is equivalent to each of the following statements.*

- *(a) The probability measure $\mathbb{Q}$ is absolutely continuous with respect to a rotation invariant measure $\overline{\mathbb{Q}}$ on $\mathcal{K}_0$.*
- *(b) For $\tilde{\mathbb{Q}}$–almost every $K \in \mathcal{K}_0$ the measure $\rho(K, \cdot)$ is absolutely continuous with respect to ν.*
- *(c) The measure μ on $\mathcal{K}_0 \times SO_d$ defined by*

$$\mu := \int_{\mathcal{K}_0} \int_{SO_d} \mathbf{1}\{(K, \vartheta) \in \cdot\} \rho(K, d\vartheta) \tilde{\mathbb{Q}}(dK)$$

is absolutely continuous with respect to $\tilde{\mathbb{Q}} \otimes \nu$.

Proof. At first we show that the rotation regularity of $\mathbb{Q}$ implies (a). For $A \in \mathcal{B}(\mathcal{K}_0)$ the decomposition (7.10), Fubini's theorem, the rotation invariance of $\tilde{\mathbb{Q}}$ and (7.11) imply

$$\mathbb{Q}(A) = \int_{SO_d} \int_{\mathcal{K}_0} \mathbf{1}\{\vartheta K \in A\} \eta(K, \vartheta) \tilde{\mathbb{Q}}(dK) \nu(d\vartheta)$$

$$= \int_{SO_d} \int_{\mathcal{K}_0} \mathbf{1}\{K \in A\} \eta(\vartheta^{-1}K, \vartheta) \tilde{\mathbb{Q}}(dK) \nu(d\vartheta)$$
$$= \int_{SO_d} \int_{\mathcal{K}_0} \mathbf{1}\{K \in A\} \eta(K, id) \tilde{\mathbb{Q}}(dK) \nu(d\vartheta)$$
$$= \int_{\mathcal{K}_0} \mathbf{1}\{K \in A\} \eta(K, id) \tilde{\mathbb{Q}}(dK),$$

which shows that $\mathbb{Q}$ is absolutely continuous with respect to the rotation invariant measure $\tilde{\mathbb{Q}}$ with Radon-Nikodym derivative $K \mapsto \eta(K, id)$.

Now we show that (a) implies (b). Let $f : \mathcal{K}_0 \to [0, \infty)$ be the Radon-Nikodym derivative of $\mathbb{Q}$ with respect to the measure $\overline{\mathbb{Q}}$. Then, we obtain for $A \in \mathcal{B}(\mathcal{K}_0)$ by the definition of $\tilde{\mathbb{Q}}$, by the absolute continuity of $\mathbb{Q}$ with respect to $\overline{\mathbb{Q}}$, by the rotation invariance of $\overline{\mathbb{Q}}$ and by Fubini's theorem that

$$\tilde{\mathbb{Q}}(A) = \int_{SO_d} \mathbb{Q}(\vartheta A) \nu(d\vartheta) = \int_{SO_d} \int_{\mathcal{K}_0} \mathbf{1}\{K \in \vartheta A\} f(K) \overline{\mathbb{Q}}(dK) \nu(d\vartheta)$$
$$= \int_{\mathcal{K}_0} \mathbf{1}\{K \in A\} \int_{SO_d} f(\vartheta K) \nu(d\vartheta) \overline{\mathbb{Q}}(dK),$$

which implies

$$\mathbb{Q} = \int_{\mathcal{K}_0} \mathbf{1}\{K \in \cdot\} \tilde{f}(K) \tilde{\mathbb{Q}}(dK)$$

with

$$\tilde{f}(K) := \mathbf{1}\left\{ \int_{SO_d} f(\vartheta K) \nu(d\vartheta) \neq 0 \right\} \left(\int_{SO_d} f(\vartheta K) \nu(d\vartheta) \right)^{-1} f(K), \quad K \in \mathcal{K}_0.$$

The rotation invariance of $\tilde{\mathbb{Q}}$ and Fubini's theorem imply for $A \in \mathcal{B}(\mathcal{K}_0)$ that

$$\mathbb{Q}(A) = \int_{\mathcal{K}_0} \mathbf{1}\{K \in A\} \tilde{f}(K) \tilde{\mathbb{Q}}(dK)$$
$$= \int_{SO_d} \int_{\mathcal{K}_0} \mathbf{1}\{\vartheta K \in A\} \tilde{f}(\vartheta K) \tilde{\mathbb{Q}}(dK) \nu(d\vartheta)$$
$$= \int_{\mathcal{K}_0} \int_{SO_d} \mathbf{1}\{\vartheta K \in A\} \tilde{f}(\vartheta K) \nu(d\vartheta) \tilde{\mathbb{Q}}(dK).$$

Therefore, the kernel ρ from Theorem 7.1.1 is by its uniqueness for $\tilde{\mathbb{Q}}$-almost all $K \in \mathcal{K}_0$ absolutely continuous with respect to ν with Radon-Nikodym derivative $\vartheta \mapsto \tilde{f}(\vartheta K)$.

Now we show that (b) implies (c). For a Borel set $E \subset \mathcal{K}_0 \times SO_d$ and $K \in \mathcal{K}_0$ let

$$E_K := \{\vartheta \in SO_d : (K, \vartheta) \in E\}.$$

Let E be a null set with respect to $\tilde{\mathbb{Q}} \otimes \nu$, then $\nu(E_K) = 0$ for $\tilde{\mathbb{Q}}$-almost every $K \in \mathcal{K}_0$. Since we assume that (b) holds, the measure $\rho(K, \cdot)$ is absolutely continuous with respect to ν. This implies $\rho(K, E_K) = 0$ for $\tilde{\mathbb{Q}}$-almost every $K \in \mathcal{K}_0$ and thus

$$\mu(E) = \int_{\mathcal{K}_0} \rho(K, E_K) \tilde{\mathbb{Q}}(dK) = 0,$$

which yields (c).

Now we show that assertion (c) implies the rotation regularity of $\mathbb{Q}$. We denote by η'' the density of μ with respect to $\tilde{\mathbb{Q}} \otimes \nu$. For every Borel set $A \subset \mathcal{K}_0$ the definition of μ implies

$$\tilde{\mathbb{Q}}(A) = \mu(A \times SO_d) = \int_A \int_{SO_d} \eta''(K, \vartheta) \nu(d\vartheta) \tilde{\mathbb{Q}}(dK). \tag{7.13}$$

Thus, there is a null set $N_1 \subset \mathcal{K}_0$ such that

$$\int_{SO_d} \eta''(K, \vartheta) \nu(d\vartheta) = 1, \quad K \in \mathcal{K}_0 \setminus N_1.$$

Let

$$\eta'(K, \vartheta) := \begin{cases} \eta''(K, \vartheta), & K \in \mathcal{K}_0 \setminus N_1 \\ 1, & K \in N_1. \end{cases}$$

Then, η' fulfils in addition to (7.13) also the property (7.11). Now, let

$$\eta(K, \vartheta) := \int_{SO_d} \eta'(\varrho K, \vartheta \varrho^{-1}) \nu(d\varrho).$$

Then, the rotation invariance of ν and Fubini's theorem imply that η fulfils (7.11) and (7.12). The representation (7.2) and the fact that $\eta'(K, \cdot)$ is for $\tilde{\mathbb{Q}}$-almost every $K \in \mathcal{K}_0$ a ν-density of $\rho(K, \cdot)$ imply for $A \in \mathcal{B}(\mathcal{K}_0)$ that

$$\mathbb{Q}(A) = \int_{\mathcal{K}_0} \int_{SO_d} \mathbf{1}\{\vartheta K \in A\} \eta'(K, \vartheta) \nu(d\vartheta) \tilde{\mathbb{Q}}(dK).$$

Thus, the rotation invariance of $\tilde{\mathbb{Q}}$ and ν and Fubini's theorem yield

$$\mathbb{Q}(A) = \int_{\mathcal{K}_0} \int_{SO_d} \mathbf{1}\{\vartheta K \in A\} \int_{SO_d} \eta'(\varrho K, \vartheta \varrho^{-1}) \nu(d\varrho) \nu(d\vartheta) \tilde{\mathbb{Q}}(dK)$$

$$= \int_{\mathcal{K}_0} \int_{SO_d} \mathbf{1}\{\vartheta K \in A\} \eta(K, \vartheta) \nu(d\vartheta) \tilde{\mathbb{Q}}(dK),$$

which implies that η fulfils (7.10) and shows thus the rotation regularity of $\mathbb{Q}$. □

We call a random convex body Z_0 rotation regular if its probability distribution is rotation regular. Of particular relevance is the notion of rotation regularity in the context of particle processes.

Definition 7.2.2. A stationary process X of convex particles in $\mathbb{R}^d$ is called rotation regular if the grain distribution $\mathbb{Q}$ has a decomposition of the form

$$\mathbb{Q}(A) = \int_{\mathcal{K}_0} \int_{SO_d} \mathbf{1}\{\vartheta K \in A\} \eta(K, \vartheta) \nu(d\vartheta) \tilde{\mathbb{Q}}(dK), \quad A \in \mathcal{B}(\mathcal{K}_0), \tag{7.14}$$

with a rotation invariant probability measure $\tilde{\mathbb{Q}}$ on $\mathcal{K}_0$ and a measurable function $\eta \geq 0$ on $\mathcal{K}_0 \times SO_d$ satisfying

$$\int_{SO_d} \eta(K, \vartheta) \nu(d\vartheta) = 1, \quad K \in \mathcal{K}_0 \tag{7.15}$$

and

$$\eta(\varrho K, \vartheta) = \eta(K, \vartheta\varrho), \quad K \in \mathcal{K}_0, \vartheta, \varrho \in SO_d. \tag{7.16}$$

If X has the intensity $\gamma > 0$, we say that X is represented by the triple $(\gamma, \eta, \tilde{\mathbb{Q}})$. This representation is unique in the sense that γ and $\tilde{\mathbb{Q}}$ are uniquely determined by X and η is uniquely determined $\tilde{\mathbb{Q}} \otimes \nu$-almost everywhere.

The decomposition Lemma 2.5.1 of the Haar measure ν on the rotation group can be used to derive a useful representation of the grain distribution $\mathbb{Q}$. Recall from Section 2.5 that we can map a unit vector $v \in S^{d-1}$ in a measurable way to a rotation $\vartheta_v \in SO_d$ which rotates the first standard basis vector e_1 to v. This allows us to replace η by a special function on the unit sphere, which is crucial for later applications of results on the polynomial approximation of functions on the unit sphere. Namely, we define for $K \in \mathcal{K}_0$ the function $\eta_K : S^{d-1} \to [0, \infty)$ by

$$\eta_K(v) = \eta(K, \vartheta_v), \quad v \in S^{d-1}.$$

Now, we obtain the following decomposition result for $\mathbb{Q}$.

Lemma 7.2.3. *Let $\mathbb{Q}$ be a rotation regular probability measure on $\mathcal{K}_0$. Then*

$$\mathbb{Q}(A) = \frac{1}{\omega_d} \int_{\mathcal{K}_0} \int_{S^{d-1}} \mathbf{1}\{\vartheta_v K \in A\} \eta_K(v) \sigma(dv) \tilde{\mathbb{Q}}(dK), \quad A \in \mathcal{B}(\mathcal{K}_0).$$

Proof. By Lemma 2.5.1 we obtain

$$\mathbb{Q}(A) = \int_{\mathcal{K}_0} \int_{SO_d} \mathbf{1}\{\vartheta K \in A\}\eta(K,\vartheta)\nu(d\vartheta)\tilde{\mathbb{Q}}(dK)$$
$$= \int_{\mathcal{K}_0} \frac{1}{\omega_d} \int_{S^{d-1}} \int_{SO_d(e_1)} \mathbf{1}\{\vartheta_v\vartheta K \in A\}\eta(K,\vartheta_v\vartheta)\nu_{e_1}(d\vartheta)\sigma(dv)\tilde{\mathbb{Q}}(dK).$$

Then, (7.16), Fubini's theorem and the rotation invariance of $\tilde{\mathbb{Q}}$ yield

$$\mathbb{Q}(A) = \int_{\mathcal{K}_0} \frac{1}{\omega_d} \int_{S^{d-1}} \int_{SO_d(e_1)} \mathbf{1}\{\vartheta_v\vartheta K \in A\}\eta(\vartheta K,\vartheta_v)\nu_{e_1}(d\vartheta)\sigma(dv)\tilde{\mathbb{Q}}(dK)$$
$$= \int_{SO_d(e_1)} \int_{\mathcal{K}_0} \frac{1}{\omega_d} \int_{S^{d-1}} \mathbf{1}\{\vartheta_v\vartheta K \in A\}\eta(\vartheta K,\vartheta_v)\sigma(dv)\tilde{\mathbb{Q}}(dK)\nu_{e_1}(d\vartheta)$$
$$= \int_{\mathcal{K}_0} \frac{1}{\omega_d} \int_{S^{d-1}} \mathbf{1}\{\vartheta_v K \in A\}\eta(K,\vartheta_v)\sigma(dv)\tilde{\mathbb{Q}}(dK)$$

and thus the assertion. □

7.3 Absolute continuity of densities of area measures

Let X be a rotation-regular particle process with convex particles represented by the triple $(\gamma, \eta, \tilde{\mathbb{Q}})$. The following lemma shows that the densities of the (mixed) area measures of X are absolutely continuous with respect to the spherical Lebesgue measure. To indicate Radon-Nikodym derivatives of the densities of the (mixed) area measures we replace the upper case psi in the notation for the measures by a minuscule psi.

Lemma 7.3.1. *Let $j \in \{0, \dots, d-1\}$, $m \in \operatorname{mix}(j,k)$, $k \in \mathbb{N}$ and $g : S^{d-1} \to \mathbb{R}$ be measurable, then*

$$\int_{S^{d-1}} g(v)\overline{\Psi}_m(X,\dots,X;dv) = \omega_d^{-1} \int_{S^{d-1}} g(v)\,\overline{\psi}_m(X,\dots,X;v)\,\sigma(dv),$$

where

$$\overline{\psi}_m(X,\dots,X;v) := \gamma^k \int_{(\mathcal{K}_0)^k} \int_{S^{d-1}} \int_{SO_d(e_1)} \prod_{i=1}^k \eta(K_i,\vartheta_v\vartheta\vartheta_u^{-1})\nu_{e_1}(d\vartheta)\Psi_m(K_1,\dots,K_k;du)$$
$$\tilde{\mathbb{Q}}^k(d(K_1,\dots,K_k)),$$

for $v \in S^{d-1}$.

Proof. By (7.14) and the rotation covariance of Ψ_m we get

$$\begin{aligned}
&\int_{S^{d-1}} g(u)\overline{\Psi}_m(X,\dots,X;du)\\
&= \gamma^k \int_{(SO_d)^k}\int_{(\mathcal{K}_0)^k}\int_{S^{d-1}} g(u)\Psi_m(\vartheta_1 K_1,\dots,\vartheta_k K_k;du)\prod_{i=1}^k \eta(K_i,\vartheta_i)\\
&\quad \tilde{\mathbb{Q}}^k(d(K_1,\dots,K_k))\nu^k(d(\vartheta_1,\dots,\vartheta_k))\\
&= \gamma^k \int_{(SO_d)^k}\int_{(\mathcal{K}_0)^k}\int_{S^{d-1}} g(\vartheta_1 u)\Psi_m(K_1,\vartheta_1^{-1}\vartheta_2 K_2,\dots,\vartheta_1^{-1}\vartheta_k K_k;du)\prod_{i=1}^k \eta(K_i,\vartheta_i)\\
&\quad \tilde{\mathbb{Q}}^k(d(K_1,\dots,K_k))\nu^k(d(\vartheta_1,\dots,\vartheta_k)).
\end{aligned} \tag{7.17}$$

By (7.16) we have $\eta(K_i,\vartheta_i) = \eta(K_i,\vartheta_1\vartheta_1^{-1}\vartheta_i) = \eta(\vartheta_1^{-1}\vartheta_i K_i,\vartheta_1)$. Applying this in (7.17) and using the rotation invariance of $\tilde{\mathbb{Q}}$ and Fubini's theorem we obtain

$$\begin{aligned}
&\int_{S^{d-1}} g(u)\overline{\Psi}_m(X,\dots,X;du)\\
&= \gamma^k \int_{(SO_d)^k}\int_{(\mathcal{K}_0)^k}\int_{S^{d-1}} g(\vartheta_1 u)\Psi_m(K_1,\vartheta_1^{-1}\vartheta_2 K_2,\dots,\vartheta_1^{-1}\vartheta_k K_k;du)\prod_{i=1}^k \eta(\vartheta_1^{-1}\vartheta_i K_i,\vartheta_1)\\
&\quad \tilde{\mathbb{Q}}^k(d(K_1,\dots,K_k))\nu^k(d(\vartheta_1,\dots,\vartheta_k))\\
&= \gamma^k \int_{(SO_d)^{k-1}}\int_{SO_d}\int_{(\mathcal{K}_0)^k}\int_{S^{d-1}} g(\vartheta_1 u)\Psi_m(K_1,K_2,\dots,K_k;du)\prod_{i=1}^k \eta(K_i,\vartheta_1)\\
&\quad \tilde{\mathbb{Q}}^k(d(K_1,\dots,K_k))\nu(d\vartheta_1)\nu^{k-1}(d(\vartheta_2,\dots,\vartheta_k))\\
&= \gamma^k \int_{(\mathcal{K}_0)^k}\int_{S^{d-1}}\int_{SO_d} g(\vartheta_1 u)\prod_{i=1}^k \eta(K_i,\vartheta_1)\nu(d\vartheta_1)\Psi_m(K_1,\dots,K_k;du)\tilde{\mathbb{Q}}^k(d(K_1,\dots,K_k)).
\end{aligned}$$

For fixed $u\in S^{d-1}$ we obtain by the rotation invariance of ν and by Lemma 2.5.1 for the inner integral that

$$\begin{aligned}
\int_{SO_d} g(\vartheta_1 u)\prod_{i=1}^k \eta(K_i,\vartheta_1)\nu(d\vartheta_1) &= \int_{SO_d} g(\vartheta_1 e_1)\prod_{i=1}^k \eta(K_i,\vartheta_1\vartheta_u^{-1})\nu(d\vartheta_1)\\
&= \omega_d^{-1}\int_{S^{d-1}}\int_{SO_d(e_1)} g(v)\prod_{i=1}^k \eta(K_i,\vartheta_v\vartheta\vartheta_u^{-1})\nu_{e_1}(d\vartheta)\sigma(dv).
\end{aligned}$$

This yields together with Fubini's theorem that

$$\int_{S^{d-1}} g(v)\overline{\Psi}_m(X,\dots,X;dv) = \omega_d^{-1} \int_{S^{d-1}} g(v)\gamma^k \int_{(\mathcal{K}_0)^k} \int_{S^{d-1}} \int_{SO_d(e_1)} \prod_{i=1}^{k} \eta(K_i, \vartheta_v \vartheta \vartheta_u^{-1}) \nu_{e_1}(d\vartheta)$$
$$\Psi_m(K_1,\dots,K_k;du)\tilde{\mathbb{Q}}^k(d(K_1,\dots,K_k))\sigma(dv)$$
$$= \omega_d^{-1} \int_{S^{d-1}} g(v)\,\overline{\psi}_m(X,\dots,X;v)\,\sigma(dv).$$

□

If X is additionally a Poisson particle process, the absolute continuity of the densities of the mixed area measures of X carries over to the densities of the area measures of the Boolean model Z associated to X.

Lemma 7.3.2. *Let $j \in \{0,\dots,d-1\}$ and $g : S^{d-1} \to \mathbb{R}$ be measurable. Then*

$$\int_{S^{d-1}} g(v)\overline{\Psi}_j(Z,dv) = \omega_d^{-1} \int_{S^{d-1}} g(v)\overline{\psi}_j(Z,v)\sigma(dv),$$

where

$$\overline{\psi}_j(Z,v) := e^{-\overline{V}_d(X)} \sum_{m\in \mathrm{mix}(j)} \frac{(-1)^{|m|-1}}{|m|!}\, \overline{\psi}_m(X,\dots,X;v), \quad v \in S^{d-1},$$

with $\overline{\psi}_m(X,\dots,X;\cdot)$ for $m \in \mathrm{mix}(j)$ defined as in Lemma 7.3.1.

Proof. By Corollary 4.1.4 and Lemma 7.3.1 we have

$$\int_{S^{d-1}} g(v)\overline{\Psi}_j(Z,dv) = e^{-\overline{V}_d(X)} \sum_{m\in \mathrm{mix}(j)} \frac{(-1)^{|m|-1}}{|m|!} \int_{S^{d-1}} g(v)\overline{\Psi}_m(X,\dots,X;dv)$$
$$= e^{-\overline{V}_d(X)} \sum_{m\in \mathrm{mix}(j)} \frac{(-1)^{|m|-1}}{|m|!} \omega_d^{-1} \int_{S^{d-1}} g(v)\,\overline{\psi}_m(X,\dots,X;v)\,\sigma(dv),$$

which implies the assertion. □

Our idea is to make use of the Hilbert space structure on $L^2(S^{d-1})$ to obtain series representations of the densities of the harmonic intrinsic volumes. Thus, an important integrability condition will be the finiteness of

$$\overline{a}(X) := \gamma \int_{\mathcal{K}_0} \eta(K,\mathrm{id})^2 V_d(K+B^d)\tilde{\mathbb{Q}}(dK) \tag{7.18}$$

since it implies $\overline{\psi}_m(X,\dots,X;\cdot) \in L^2(S^{d-1})$ by the following lemma.

Lemma 7.3.3. *Let $j \in \{0,\dots,d-1\}, k \in \mathbb{N}$ and $m \in \operatorname{mix}(j,k)$. Then*

$$\|\overline{\psi}_m(X,\dots,X;\cdot)\|_2^2 \le c_d\, \overline{a}(X)^k \prod_{i=1}^k \overline{V}_{m_i}(X),$$

with some $c_d > 0$ which depends only on d.

Proof. For $K_1,\dots,K_k \in \mathcal{K}$ the decomposability property Theorem 3.2.2,(ii) and the translative integral formula Theorem 3.2.1 can be applied to the mixed functional $V_m(K_1,\dots,K_k)$, which arises as the full support measure. This yields together with the monotonicity of V_j the bound

$$\begin{aligned}
&V_m(K_1,\dots,K_k)\\
&= \kappa_d^{-1} V_{d,m}(B^d, K_1,\dots,K_k)\\
&\le \kappa_d^{-1} \int\limits_{(\mathbb{R}^d)^k} V_j(B^d \cap (K_1+x_1)\cap\ldots\cap(K_k+x_k))d(x_1,\dots,x_k)\\
&\le \kappa_d^{-1} V_j(B^d) \int\limits_{(\mathbb{R}^d)^k} \mathbf{1}\{B^d \cap (K_1+x_1)\cap\ldots\cap(K_k+x_k) \neq \emptyset\} d(x_1,\dots,dx_k)\\
&\le \kappa_d^{-1} V_j(B^d) \prod_{i=1}^k V_d(K_i + B^d).
\end{aligned} \tag{7.19}$$

Furthermore, the rotational formula [SW08, (6.20)] implies

$$\int\limits_{(SO_d)^{k-1}} V_m(K_1, \vartheta_2 K_2,\dots,\vartheta_k K_k)\nu^{k-1}(d(\vartheta_2,\dots,\vartheta_k)) \le c \prod_{i=1}^k V_{m_i}(K_i), \tag{7.20}$$

for some $c > 0$ depending only on d. By Lemma 7.3.1 and Hölder's inequality we get

$$\begin{aligned}
&\int\limits_{S^{d-1}} \overline{\psi}_m(X,\dots,X;v)^2 \sigma(dv)\\
&= \int\limits_{S^{d-1}} \Big(\gamma^k \int\limits_{(\mathcal{K}_0)^k} \int\limits_{S^{d-1}} \int\limits_{SO_d(e_1)} \prod_{i=1}^k \eta(K_i, \vartheta_v \vartheta \vartheta_u^{-1}) \nu_{e_1}(d\vartheta) \Psi_m(K_1,\dots,K_k;du)\\
&\quad \times \tilde{\mathbb{Q}}^k(d(K_1,\dots,K_k))\Big)^2 \sigma(dv)\\
&\le \int\limits_{S^{d-1}} \gamma^k \int\limits_{(\mathcal{K}_0)^k} \int\limits_{S^{d-1}} \int\limits_{SO_d(e_1)} \prod_{i=1}^k \eta(K_i, \vartheta_v \vartheta \vartheta_u^{-1})^2 \nu_{e_1}(d\vartheta) \Psi_m(K_1,\dots,K_k;du)
\end{aligned}$$

$$\times\, \tilde{\mathbb{Q}}^k(d(K_1,\dots,K_k))\sigma(dv)\gamma^k \int\limits_{(\mathcal{K}_0)^k} V_m(K_1,\dots,K_k)\tilde{\mathbb{Q}}^k(d(K_1,\dots,K_k)). \tag{7.21}$$

Then, Fubini's theorem, Lemma 2.5.1, the rotation invariance of ν, the upper bound (7.19) for the mixed functional V_m, the shift property (7.16) and the rotation invariance of the volume of the parallel set and of the measure $\tilde{Q}$ yield for the first multiple integral in (7.21) that

$$\begin{aligned}
&\gamma^k \int\limits_{S^{d-1}} \int\limits_{(\mathcal{K}_0)^k} \int\limits_{S^{d-1}} \int\limits_{SO_d(e_1)} \prod_{i=1}^k \eta(K_i, \vartheta_v\vartheta\vartheta_u^{-1})^2 \nu_{e_1}(d\vartheta)\Psi_m(K_1,\dots,K_k;du) \\
&\tilde{\mathbb{Q}}^k(d(K_1,\dots,K_k))\sigma(dv) \\
&= \omega_d\,\gamma^k \int\limits_{(\mathcal{K}_0)^k} \int\limits_{S^{d-1}} \int\limits_{SO_d} \prod_{i=1}^k \eta(K_i, \vartheta\vartheta_u^{-1})^2 \nu(d\vartheta)\Psi_m(K_1,\dots,K_k;du)\tilde{\mathbb{Q}}^k(d(K_1,\dots,K_k)) \\
&= \omega_d\,\gamma^k \int\limits_{(\mathcal{K}_0)^k} \int\limits_{SO_d} \prod_{i=1}^k \eta(K_i, \vartheta)^2 \nu(d\vartheta) V_m(K_1,\dots,K_k)\tilde{\mathbb{Q}}^k(d(K_1,\dots,K_k)) \\
&\le c\,\gamma^k \int\limits_{SO_d} \prod_{i=1}^k \int\limits_{\mathcal{K}_0} \eta(K_i,\vartheta)^2 V_d(K_i + B^d)\tilde{\mathbb{Q}}(dK_i)\nu(d\vartheta) \\
&= c\,\gamma^k \int\limits_{SO_d} \left(\int\limits_{\mathcal{K}_0} \eta(\vartheta K, \mathrm{id})^2 V_d(\vartheta K + B^d)\tilde{\mathbb{Q}}(dK) \right)^k \nu(d\vartheta) &(7.22)\\
&= c\,\gamma^k \left(\int\limits_{\mathcal{K}_0} \eta(K, \mathrm{id})^2 V_d(K + B^d)\tilde{\mathbb{Q}}(dK) \right)^k &(7.23)\\
&= c\,\overline{a}(X)^k, &(7.24)
\end{aligned}$$

with some $c > 0$ depending only on d. For the remaining factor in (7.21) the rotation invariance of $\tilde{\mathbb{Q}}$ and (7.20) imply

$$\gamma^k \int\limits_{(\mathcal{K}_0)^k} V_m(K_1,\dots,K_k)\tilde{\mathbb{Q}}^k(d(K_1,\dots,K_k)) \le c \prod_{i=1}^k \overline{V}_{m_i}(X), \tag{7.25}$$

with some $c > 0$ depending only on d. Inserting (7.24) and (7.25) in (7.21) we obtain the assertion. □

The integrability condition (7.18) ensures also the L^2-integrability of the densities of the area measures of the Boolean model associated to X.

Lemma 7.3.4. *Let* $j \in \{0, \ldots, d-1\}$. *Then*

$$\|\overline{\psi}_j(Z;\cdot)\|_2^2 \le c_d\, e^{-2\overline{V}_d(X)} \sum_{m \in \operatorname{mix}(j)} \overline{a}(X)^{|m|} \prod_{i=1}^{|m|} \overline{V}_{m_i}(X)$$

with some $c_d > 0$ *which depends only on d.*

Proof. The representation of $\overline{\psi}_j(Z,\cdot)$ derived in Lemma 7.3.2 implies

$$\|\overline{\psi}_j(Z,\cdot)\|_2^2 = e^{-2\overline{V}_d(X)} \omega_d^{-1} \int_{S^{d-1}} \left(\sum_{m \in \operatorname{mix}(j)} \frac{(-1)^{|m|-1}}{|m|!} \overline{\psi}_m(X,\ldots,X;v) \right)^2 \sigma(dv).$$

For each $v \in S^{d-1}$ Hölder's inequality implies

$$\begin{aligned} &\left(\sum_{m \in \operatorname{mix}(j)} \frac{(-1)^{|m|-1}}{|m|!} \overline{\psi}_m(X,\ldots,X;v) \right)^2 \\ &\le \sum_{m \in \operatorname{mix}(j)} (|m|!)^{-2} \sum_{m \in \operatorname{mix}(j)} \overline{\psi}_m(X,\ldots,X;v)^2. \end{aligned}$$

This and the upper bounds for the L^2-norm of the densities of the mixed area measures of X obtained in Lemma 7.3.3 yields the assertion. □

7.4 A modulus of isotropy for rotation regular particle processes

In this subsection we give first a motivation for the introduction of a modulus of isotropy and describe its construction in the planar case. Then, we give a short overview of different moduli of smoothness on the unit sphere, in particular we present the recently introduced modulus of smoothness on the unit sphere by Dai and Xu [DX10] and its relation to approximation of functions on the unit sphere by spherical polynomials. We define a modulus of smoothness on the rotation group and investigate its relation to the modulus of smoothness on the unit sphere. Then, we introduce a modulus of isotropy for a rotation regular particle process. This quantity measures the global variation of the grain distribution with respect to rotations by angles smaller than a prescribed parameter and will play an important role for the speed of convergence of series of densities of geometric functionals related to a particle process.

7.4.1 Motivation

In applications isotropy is almost never given. Since statistical methods are often developed under this assumption, small deviations from isotropy have to be neglected. It is therefore desirable to define a global quantity for the degree of anisotropy and to relate it to the error made when applying methods which are designed for the isotropic situation. In stochastic geometry the method of densities for the estimation of the intensity from the densities of the intrinsic volumes of a stationary Boolean model is developed under the assumption of isotropy. Our aim is to show that the method of densities is also applicable in the non-isotropic situation if the modulus of isotropy is small. To provide the reader with some intuition, we describe now the basic construction of this modulus of isotropy in the planar case.

At first we assume that the typical grain Z_0 in the plane is obtained by rotating a fixed base grain $E \in \mathcal{K}_0$ by a random angle $\theta \in [0, 2\pi)$ with probability density f. We extend f to a 2π-periodic function on the real line. Then, the traditional definition of the modulus of continuity going back to Laplace is

$$\omega(f,t) := \sup_{|x-y|\leq t} |f(x) - f(y)|, \; t > 0. \tag{7.26}$$

We can interpret the value $\omega(f,t)$ as a measure of the degree of anisotropy with respect to rotations by angles smaller than t.

Now, assume that E is randomly chosen from $\mathcal{K}_0$ according to a rotation invariant probability measure $\tilde{\mathbb{Q}}$ and then rotated by a random angle with probability density f_E depending on the realization E. In view of the disintegration result from the last section this is equivalent to a minor assumption of regularity on the distribution of the typical grain. Now we may consider the value

$$\int_{\mathcal{K}_0} \omega(f_E, t)\tilde{\mathbb{Q}}(dE), \quad t \geq 0$$

as a global measure of the deviation from isotropy. Since the method of densities relies on the densities of the intrinsic volumes it will be of advantage to weight not only probabilistically but also with respect to the volume of the parallel set of E, which leads us to the integral

$$\int_{\mathcal{K}_0} \omega(f_E, t) V_2\left(E + B^2\right) \tilde{\mathbb{Q}}(dE), \quad t \geq 0$$

as a measure of anisotropy. Another important motivation of the above construction is that the modulus of smoothness is related to the error made when approximating a L^2-integrable 2π-periodic function by trigonometric polynomials.

7.4.2 A modulus of smoothness on the unit sphere

There exist different attempts to define convenient moduli of continuity on the unit sphere in dimensions higher than two. Calderón, Weiss and Zygmund [CWZ67] replace the distance $|x-y|$, $x, y \in \mathbb{R}$ in (7.26) by the geodesic distance $d(u,v) := \arccos\langle u, v\rangle$, $u, v \in S^{d-1}$. Based on rotations one can define also moduli of smoothness of higher order. For $\vartheta \in SO_d$ we define the rotation operator T_ϑ applied to a function $f : S^{d-1} \to \mathbb{R}$ as

$$[T_\vartheta f](u) := f(\vartheta u), \quad u \in S^{d-1}$$

and the difference operator Δ_ϑ by

$$\Delta_\vartheta := T_\vartheta - I.$$

For $r \in \mathbb{N}$ the r-fold iterated difference operator is

$$\Delta_\vartheta^r = (T_\vartheta - I)^r = \sum_{q=0}^{r} \binom{r}{q} (-1)^{r-q} T_\vartheta^q. \tag{7.27}$$

We define a metric d on SO_d by

$$d(\vartheta_1, \vartheta_2) := \max_{u \in S^{d-1}} d(\vartheta_1 u, \vartheta_2 u), \quad \vartheta_1, \vartheta_2 \in SO_d,$$

where d denotes on the unit sphere the geodesic distance and we need the set of rotations

$$SO_d(t) := \{\vartheta \in SO_d : d(\vartheta, id) \le t\}, \quad t > 0,$$

which have distance at most t from the identity rotation. A modulus of smoothness of order $r \in \mathbb{N}$ (or modulus of continuity for $r = 1$) based on rotations was introduced by Ditzian [Dit99] for measurable functions $f : S^{d-1} \to \mathbb{R}$ as

$$\tilde{\omega}_r(f,t)_p := \sup_{\vartheta \in SO_d(t)} \|\Delta_\vartheta^r f\|_p, \quad t > 0, p \in [1, \infty]. \tag{7.28}$$

Observe that for $r = 1$ and $p = \infty$ the definition (7.28) coincides with the one in [CWZ67], which is based on geodesic distances. In the following we use also the recently introduced modulus of smoothness by Dai and Xu [DX10], which is obtained if we replace the set $SO_d(t)$ in (7.28) by the subset of $SO_d(t)$ which contains only rotations in planes which are spanned by two coordinate axes. It has the advantage that it can be computed more easily because only angles in finitely many planes are considered.

For a more complete overview of the different moduli of smoothness on the unit sphere we refer the reader to the introduction of [DX10] and to the relevant chapters in two recent monographs on approximation theory on the unit sphere, [AH12, Chapter 4], which presents the three-dimensional case, and [DX13, Chapter 4].

Now, we introduce the modulus of smoothness by Dai and Xu in more detail. Let $\theta \in \mathbb{R}$ and $1 \leq i < j \leq d$. We denote by $\vartheta_{i,j,\theta} \in SO_d$ the rotation in the $(i,j)-$plane by the angle θ. For $x = (x_1, \ldots, x_d)^\top \in \mathbb{R}^d$ this means for example

$$\vartheta_{1,2,\theta}\, x = (s\cos(\varphi + \theta), s\sin(\varphi + \theta), x_3, \ldots, x_d),$$

if $(x_1, x_2) = s(\cos\varphi, \sin\varphi)$ for $s > 0, \varphi \in [0, 2\pi)$. We use the abbreviation

$$\Delta_{i,j,\theta} := \Delta_{\vartheta_{i,j,\theta}}.$$

For measurable $f : S^{d-1} \to \mathbb{R}$ Dai and Xu's modulus of smoothness of order $r \in \mathbb{N}$ with respect to the L^p-norm with $p \in [1, \infty]$ is

$$\omega_r(f,t)_p := \max_{1 \leq i < j \leq d} \sup_{|\theta| \leq t} \|\Delta^r_{i,j,\theta} f\|_p, \quad t > 0. \tag{7.29}$$

For the modulus of smoothness of first order we omit the index r. In most cases we choose $p = 2$, in this case we omit the index p. Hence, we use the abbreviations

$$\omega(f,t) := \omega_1(f,t)_2, \quad \omega(f,t)_p := \omega_1(f,t)_p \text{ and } \omega_r(f,t) := \omega_r(f,t)_2.$$

We use the same notational conventions with respect to r and p for the modulus of continuity defined in (7.28).

In [DX10] the modulus of smoothness $\omega_r(f,t)_p$ is defined only for functions $f \in L^p(S^{d-1})$, $1 \leq p < \infty$, or $f \in C(S^{d-1})$ for $p = \infty$. For our purposes it is more convenient to drop these assumptions and to allow the modulus of smoothness to assume the value ∞. The moduli of smoothness $\omega_r(f,t)_p$ and $\tilde{\omega}_r(f,t)_p$ are closely related to each other. In two dimensions the definitions of both moduli of smoothness coincide. Thus, we obtain for all measurable $f : S^1 \to \mathbb{R}$ that

$$\omega_r(f,t)_p = \tilde{\omega}_r(f,t)_p. \tag{7.30}$$

In arbitrary dimension $d \geq 2$ one obtains the following relation.

Proposition 7.4.1. *Let $r \in \mathbb{N}$, $p \in [1, \infty]$, $t > 0$ and $f : S^{d-1} \to \mathbb{R}$ be a measurable function. Then*

$$\omega_r(f,t)_p \leq \tilde{\omega}_r(f,t)_p \tag{7.31}$$

and

$$\tilde{\omega}(f,t) \leq c\, \omega(f,t) \tag{7.32}$$

with some constant $c > 0$ depending only on d.

Proof. A small calculation (carried out in [DX10, p.1241]) shows $\vartheta_{i,j,\theta} \in SO_d(t)$ for $0 \leq \theta \leq t$ and each $1 \leq i < j \leq d$. This implies the first inequality. The latter inequality is shown in [DX10, Corollary 3.11]. □

Remark 7.4.2. Under additional regularity assumptions on the function $f : S^{d-1} \to \mathbb{R}$ we obtain bounds for the moduli of smoothness.

The difference operator $\Delta_{i,j,\theta}$ is related to the angular derivatives $D_{i,j}$, which are defined as derivatives with respect to the angles in the (i, j)-plane. For $f \in C^1(S^{d-1})$, $x \in S^{d-1}$ with $(x_1, x_2) = s(\cos\theta, \sin\theta)$ for $s \geq 0$ and $\theta \in [0, 2\pi)$ we have

$$D_{1,2}f(x) = \frac{\partial}{\partial\theta} f(s\cos(\theta), s\sin(\theta), x_3, \ldots, x_n).$$

For $f \in C^r(S^{d-1})$ the iterated difference operator satisfies the bound

$$\|\Delta^r_{i,j,\theta} f\|_p \leq c|\theta|^r \|D^r_{i,j} f\|_p \tag{7.33}$$

with some constant $c > 0$, see [DX10, Lemma 2.6, (ii)]. This yields

$$\omega_r(f, t)_p \leq c\,t^r \max_{1\leq i<j\leq d} \|D^r_{i,j} f\|_p$$

for $f \in C^r(S^{d-1})$ and some constant $c > 0$.

If f is Lipschitz continuous with constant $L > 0$, i.e.

$$|f(u) - f(v)| \leq L\, d(u, v), \quad u, v \in S^{d-1},$$

we obtain

$$|f(\vartheta u) - f(u)| \leq L\, d(\vartheta u, u) \leq L t, \quad u \in S^{d-1}, \vartheta \in SO_d(t)$$

and thus

$$\omega(f, t) \leq \tilde{\omega}(f, t) \leq L t, \quad t > 0.$$

If f is uniformly continuous, there exists for every $\epsilon > 0$ a $\delta(\epsilon) > 0$ such that

$$|f(u) - f(v)| \leq \epsilon, \text{ if } d(u, v) \leq \delta(\epsilon), \quad u, v \in S^{d-1}.$$

For the modulus of continuity we obtain

$$\omega(f, t) \leq \tilde{\omega}(f, t) \leq \epsilon, \text{ if } t \leq \delta(\epsilon).$$

An important aspect of the modulus of smoothness is its relation to approximation on the unit sphere by spherical polynomials. Suppose we need for $f \in L^2(S^{d-1})$ an upper bound for the error

$$\|f - \pi_n f\|_2 = \inf_{q\in\Pi_n} \|f - q\|_2$$

which we make if we approximate f in an optimal way by spherical polynomials of degree at most n. Finding such bounds is a problem of approximation theory on the unit sphere [AH12, DX13]. The error of best approximation can be bounded in terms of the modulus of smoothness of the function f by the following result, which was obtaind in [DX10, Theorem 3.4].

Theorem 7.4.3. *(Dai and Xu 2010) Let $n \in \mathbb{N}_0$ and $r \in \mathbb{N}$. If $f \in L^p(S^{d-1})$ and $1 \leq p < \infty$ or $f \in C(S^{d-1})$ and $p = \infty$, then*

$$\inf_{q \in \Pi_n(S^{d-1})} \|f - q\|_p \leq c\, \omega_r \left(f, \frac{1}{n+1}\right)_p,$$

with a constant $c > 0$ depending only on d and r.

7.4.3 A modulus of smoothness on the rotation group

Now we define a modulus of smoothness for functions on the rotation group in a similar way as in the last section for functions on the unit sphere.

For $\varrho \in SO_d$ we define the operators T^ϱ respectively T_ϱ operating on a function $f : SO_d \to \mathbb{R}$ by

$$[T_\varrho f](\vartheta) := f(\varrho\vartheta) \text{ and } [T^\varrho f](\vartheta) := f(\vartheta\,\varrho), \quad \vartheta \in SO_d.$$

As before let $\theta \in \mathbb{R}$ and $1 \leq i < j \leq d$. We define the difference operators Δ_ϱ and $\Delta_{i,j,\theta}$ by

$$\Delta_\varrho := T_\varrho - I$$

and

$$\Delta_{i,j,\theta} := \Delta_{\vartheta_{i,j,\theta}}.$$

For the r-fold iterated operators it holds a binomial relation

$$\Delta_\varrho^r = \sum_{q=0}^{r} \binom{r}{q} (-1)^{r-q} (T_\varrho)^q. \tag{7.34}$$

Let $r \in \mathbb{N}$, $p \in [1, \infty]$, $t > 0$ and $f : SO_d \to \mathbb{R}$ be a measurable function. Then, we define moduli of smoothness of order r by

$$\tilde{\omega}_r(f,t)_p := \sup_{\varrho \in SO_d(t)} \|\Delta_\varrho^r f\|_p \tag{7.35}$$

and

$$\omega_r(f,t)_p := \max_{1 \leq i < j \leq d} \sup_{|\theta| \leq t} \|\Delta_{i,j,\theta}^r f\|_p. \tag{7.36}$$

We use the same notational conventions with respect to r and p as for the moduli of smoothness for functions on the unit sphere.

Remark 7.4.4. As mentioned in the proof of Proposition 7.4.1 it is shown in [DX10, p.1241]) that $\vartheta_{i,j,\theta} \in SO_d(t)$ for $0 \le \theta \le t$ and each $1 \le i < j \le d$. Thus, we have

$$\omega(f,t) \le \tilde{\omega}(f,t), \quad t > 0$$

for all measurable functions $f : SO_d \to \mathbb{R}$. Similarly as in Remark 7.4.2 we obtain under additional regularity assumptions on the function $f : SO_d \to \mathbb{R}$ bounds for the moduli of smoothness of first order on the rotation group. If f is Lipschitz continuous with constant $L > 0$, i.e.

$$|f(\vartheta_1) - f(\vartheta_2)| \le L\, d(\vartheta_1, \vartheta_2), \quad \vartheta_1, \vartheta_2 \in SO_d,$$

we obtain

$$\omega(f,t) \le \tilde{\omega}(f,t) \le L t, \quad t > 0.$$

If f is uniformly continuous, there exists for every $\epsilon > 0$ a $\delta(\epsilon) > 0$ such that

$$|f(\vartheta_1) - f(\vartheta_2)| \le \epsilon, \text{ if } d(\vartheta_1, \vartheta_2) \le \delta(\epsilon), \quad \vartheta_1, \vartheta_2 \in SO_d.$$

Therefore, we obtain for the modulus of continuity that

$$\omega(f,t) \le \tilde{\omega}(f,t) \le \epsilon, \text{ if } t \le \delta(\epsilon).$$

The moduli of smoothness are invariant under the operator T^ϱ by the following lemma.

Lemma 7.4.5. *For measurable $f : SO_d \to \mathbb{R}, \varrho \in SO_d, r \in \mathbb{N}, p \in [1, \infty]$ and $t > 0$, we have*

$$\tilde{\omega}_r(f,t)_p = \tilde{\omega}_r(T^\varrho f, t)_p$$

and

$$\omega_r(f,t)_p = \omega_r(T^\varrho f, t)_p.$$

Proof. At first let $p \in [1, \infty)$. Then, we obtain by the binomial relation (7.34) and the rotation invariance of ν for $\zeta \in SO_d$ that

$$\begin{aligned} \|\Delta_\zeta^r T^\varrho f\|_p^p &= \int\limits_{SO_d} \left| \sum_{q=0}^{r} \binom{r}{q} (-1)^q f(\zeta^q \,\vartheta\, \varrho) \right|^p \nu(d\vartheta) \\ &= \int\limits_{SO_d} \left| \sum_{q=0}^{r} \binom{r}{q} (-1)^q f(\zeta^q \,\vartheta) \right|^p \nu(d\vartheta) = \|\Delta_\zeta^r f\|_p^p, \end{aligned}$$

which implies the assertion. For $p = \infty$, the assertion follows since we obtain for $\zeta \in SO_d$ that

$$\begin{aligned} \|\Delta_\zeta^r T^\varrho f\|_\infty &= \sup_{\vartheta \in SO_d} \left| \sum_{q=0}^{r} \binom{r}{q} (-1)^q f(\zeta^q \,\vartheta\, \varrho) \right| = \sup_{\vartheta \in SO_d} \left| \sum_{q=0}^{r} \binom{r}{q} (-1)^q f(\zeta^q \,\vartheta) \right| \\ &= \|\Delta_\zeta^r f\|_\infty. \end{aligned}$$

□

The moduli of smoothness of order 1 of products of functions can be bounded from above in terms of the moduli of smoothness of each function by the following lemma.

Lemma 7.4.6. *Let $f, g \in L^\infty(S^{d-1})$, $p \in [1,\infty]$ and $t > 0$. Then*

$$\tilde{\omega}(fg,t)_p \leq \tilde{\omega}(f,t)\,\|g\|_\infty + \|f\|_\infty\,\tilde{\omega}(g,t)$$

and

$$\omega(fg,t)_p \leq \omega(f,t)\,\|g\|_\infty + \|f\|_\infty\,\omega(g,t).$$

Proof. By definition we have

$$\tilde{\omega}(fg,t)_p = \sup_{\varrho \in SO_d(t)} \|T_\varrho(fg) - fg\|_p$$

and

$$\omega(fg,t)_p = \max_{1\leq i<j\leq d} \sup_{|\theta|\leq t} \left\|T_{\vartheta_{i,j,\theta}}(fg) - fg\right\|_p.$$

This implies the assertion since we get for each $\varrho \in SO_d$ that

$$\begin{aligned}\|T_\varrho(fg) - fg\|_p &= \|T_\varrho(f)\,T_\varrho(g) - fg\|_p\\ &= \|T_\varrho(f)\,T_\varrho(g) - f\,T_\varrho(g)\,\|_p + \|f\,T_\varrho(g) - fg\|_p\\ &\leq \|T_\varrho f - f\|_p \|T_\varrho g\|_\infty + \|f\|_\infty \|T_\varrho g - g\|_p\\ &= \|\Delta_\varrho f\|_p\,\|g\|_\infty + \|f\|_\infty\,\|\Delta_\varrho g\|_p.\end{aligned}$$

□

Related to a measurable function $f : SO_d \to \mathbb{R}$ and $\vartheta \in SO_d$ we define two real-valued functions f_ϑ and f^* on the unit sphere by

$$f_\vartheta(v) = f(\vartheta_v \vartheta), \quad v \in S^{d-1}$$

and

$$f^*(v) = \int\limits_{SO_d(e_1)} f_\vartheta(v)\nu_{e_1}(d\vartheta), \quad v \in S^{d-1}. \tag{7.37}$$

The map f^* has a probabilistic interpretation. Namely, $f^*(v)$ is the mean value of f on the set of rotations $SO_d(e_1, v)$, which rotate e_1 to v, with respect to the unique $SO_d(v)$-invariant probability measure on $SO_d(e_1, v)$, which we denote by $\nu_{e_1,v}$, $v \in S^{d-1}$. That is, we have

$$f^*(v) = \int\limits_{SO_d(e_1,v)} f(\vartheta)\nu_{e_1,v}(d\vartheta), \quad v \in S^{d-1}.$$

The moduli of smoothness for functions on the rotation group are related to the moduli of smoothness for functions on the unit sphere which are defined via rotations in the following way.

Lemma 7.4.7. *Let $t > 0, r \in \mathbb{N}, p \in [1,\infty]$ and $f : SO_d \to \mathbb{R}$ be measurable, then*

$$\omega_r(f^*, t)_p \leq \omega_d^{1/p}\, \omega_r(f,t)_p, \tag{7.38}$$

where one should read $1/\infty = 0$. The bound remains true if $\omega_r(f,t)_p$ is replaced by $\tilde{\omega}_r(f,t)_p$.

Proof. For all $p \in [1,\infty]$ the definition of the moduli of smoothness in (7.35) is

$$\tilde{\omega}_r(f^*, t)_p = \sup_{\zeta \in SO_d(t)} \|\Delta_\zeta^r f^*\|_p$$

respectively in (7.36) it is

$$\omega_r(f^*, t)_p = \max_{1 \leq i < j \leq d} \sup_{|\theta| \leq t} \|\Delta_{\vartheta_{i,j,\theta}}^r f^*\|_p.$$

Thus, it is sufficient to show for $\zeta \in SO_d$ that

$$\|\Delta_\zeta^r f^*\|_p \leq \omega_d^{1/p}\, \|\Delta_\zeta^r f\|_p, \quad p \in [1,\infty)$$

respectively

$$\|\Delta_\zeta^r f^*\|_\infty \leq \|\Delta_\zeta^r f\|_\infty.$$

For $v \in S^{d-1}$ it follows by the binomial relation (7.34) that

$$\left[\Delta_\zeta^r f^*\right](v) = \sum_{q=0}^{r} \binom{r}{q} (-1)^{r-q} f^*(\zeta^q v), \quad v \in S^{d-1}. \tag{7.39}$$

Let

$$v_q := \zeta^q v, \quad q \in \{0,\dots,r\}, v \in S^{d-1}.$$

Then, (7.39) and the definition of f^* imply

$$\left[\Delta_\zeta^r f^*\right](v) = \sum_{q=0}^{r} \binom{r}{q} (-1)^{r-q} \int_{SO_d(e_1)} f(\vartheta_{v_q} \vartheta) \nu_{e_1}(d\vartheta). \tag{7.40}$$

Let $q \in \{0,\dots,r\}$ and $v \in S^{d-1}$ be such that the vectors $v_q, e_2, \dots, e_d$ are linearly independent. It is sufficient to consider this situation, because the linear independence holds for σ-almost every $v \in S^{d-1}$. With the abbreviations

$$\vartheta_* := \zeta^q \vartheta_v \text{ and } \vartheta_{**} := GS(e_1, \vartheta_*^{-1} e_2, \dots, \vartheta_*^{-1} e_d) \in SO_d(e_1)$$

we obtain by (2.17), since $v_q = \vartheta_* e_1$ and by (2.16) that

$$\begin{aligned}\vartheta_{v_q} &= GS(v_q, e_2, \dots, e_d) = \zeta^q\, \vartheta_v GS(e_1, \vartheta_*^{-1} e_2, \dots, \vartheta_*^{-1} e_d) \\ &= \zeta^q\, \vartheta_v \vartheta_{**}. \end{aligned} \tag{7.41}$$

Using this relation in (7.40), we can apply the invariance property of ν_{e_1} since $\vartheta_{**} \in SO_d(e_1)$ to obtain

$$[\Delta_\zeta^r f^*](v) = \sum_{q=0}^{r} \binom{r}{q} (-1)^{r-q} \int\limits_{SO_d(e_1)} f(\zeta^q \vartheta_v \vartheta) \nu_{e_1}(d\vartheta) = \int\limits_{SO_d(e_1)} [\Delta_\zeta^r f](\vartheta_v \vartheta) \nu_{e_1}(d\vartheta).$$

For $p \in [1,\infty)$ Jensen's inequality and Lemma 2.5.1 imply

$$\|\Delta_\zeta^r f^*\|_p^p \leq \int\limits_{S^{d-1}} \int\limits_{SO_d(e_1)} |[\Delta_\zeta^r f](\vartheta_v \vartheta)|^p \nu_{e_1}(d\vartheta) \sigma(dv) = \omega_d \|\Delta_\zeta^r f\|_p^p.$$

This yields the assertion for $p \in [1,\infty)$. For $p = \infty$ the inequality follows from

$$\|\Delta_\zeta^r f^*\|_\infty = \sup_{v \in S^{d-1}} \left| \int\limits_{SO_d(e_1)} [\Delta_\zeta^r f](\vartheta_v \vartheta) \nu_{e_1}(d\vartheta) \right| \leq \sup_{\vartheta \in SO_d} |[\Delta_\zeta^r f](\vartheta)| = \|\Delta_\zeta^r f\|_\infty.$$

□

As a corollary of Theorem 7.4.3 and Lemma 7.4.7 we obtain a bound for the error of best approximation of f^*.

Corollary 7.4.8. *Let $n \in \mathbb{N}_0$ and $r \in \mathbb{N}$. If $f \in L^p(SO_d)$ and $1 \leq p < \infty$ or $f \in C(SO_d)$ and $p = \infty$, then*

$$\inf_{q \in \Pi_n(S^{d-1})} \|f^* - q\|_p \leq c\, \omega_d^{1/p}\, \omega_r\left(f, \frac{1}{n+1}\right)_p$$

where $c > 0$ depends only on r and d and with the convention $1/\infty = 0$.

7.4.4 A modulus of isotropy for a rotation regular particle process

Let X be a rotation regular particle process represented by the triple $(\gamma, \eta, \tilde{\mathbb{Q}})$ as introduced in Section 7.2. Recall that $\tilde{\mathbb{Q}}$ is a rotation invariant probability measure on $\mathcal{K}_0$ and η is a measurable function on $\mathcal{K}_0 \times SO_d$ which is for each $K \in \mathcal{K}_0$ a probability density function with respect to the Haar measure on SO_d and satisfies a special shift property. Furthermore, the grain distribution $\mathbb{Q}$ of X has by (7.14) and by Lemma 7.2.3 the two representations

$$\mathbb{Q}(A) = \int\limits_{\mathcal{K}_0} \int\limits_{SO_d} \mathbf{1}_A(\vartheta K) \eta(K,\vartheta) \nu(d\vartheta) \tilde{\mathbb{Q}}(dK) = \frac{1}{\omega_d} \int\limits_{\mathcal{K}_0} \int\limits_{S^{d-1}} \mathbf{1}_A(\vartheta_v K) \eta_K(v) \sigma(dv) \tilde{\mathbb{Q}}(dK),$$

for $A \in \mathcal{B}(\mathcal{K}_0)$. An important role will be played by the application of the $*$-operator from (7.37) to the second component of the kernel η. Thus, it is convenient to introduce the abbreviation

$$\eta_K^* := [\eta(K,\cdot)]^*, \quad K \in \mathcal{K}_0. \tag{7.42}$$

We obtain an alternative representation for η_K^*.

Lemma 7.4.9. *Let $K \in \mathcal{K}_0$. Then*

$$\eta_K^*(v) = \int\limits_{SO_d(e_1)} \eta_{\vartheta K}(v)\nu_{e_1}(d\vartheta), \quad v \in S^{d-1}.$$

Proof. By the definition (7.37) of the $*$-operator it holds

$$\eta_K^*(v) = \int\limits_{SO_d(e_1)} [\eta(K,\cdot)]_\vartheta (v)\nu_{e_1}(d\vartheta).$$

The shift property (7.16) yields

$$[\eta(K,\cdot)]_\vartheta (v) = \eta(K, \vartheta_v\vartheta) = \eta(\vartheta K, \vartheta_v) = \eta_{\vartheta K}(v). \tag{7.43}$$

This implies the assertion. □

As a consequence of Corollary 7.4.8 we obtain the following upper bound for the error of best approximation of η_K^* by spherical polynomials.

Corollary 7.4.10. *Let $n \in \mathbb{N}_0$, $r \in \mathbb{N}$ and $K \in \mathcal{K}_0$. If $\eta(K,\cdot) \in L^p(SO_d)$ and $1 \le p < \infty$ or $\eta(K,\cdot) \in C(SO_d)$ and $p = \infty$, then*

$$\inf_{q\in\Pi_n(S^{d-1})} \|\eta_K^* - q\|_p \le c\,\omega_d^{1/p}\,\omega_r\left(\eta(K,\cdot), \frac{1}{n+1}\right)_p$$

with the convention $1/\infty = 0$ and a constant $c > 0$ depending only on d and r.

Now, we introduce a modulus of isotropy for particle processes, which will play an important role for the speed of convergence of certain series of densities of geometric functionals.

Definition 7.4.11. Let $t > 0, 1 \le p \le \infty, r \in \mathbb{N}$ and X be a rotation regular particle process represented by the triple $(\gamma, \eta, \tilde{\mathbb{Q}})$. Then, we define the *modulus of isotropy* of X by

$$\overline{\omega}_r(X,t)_p := \left(\gamma \int\limits_{\mathcal{K}_0} \omega_r(\eta(K,\cdot),t)_p^2\, V_d(K+B^d)\tilde{\mathbb{Q}}(dK)\right)^{1/2}. \tag{7.44}$$

Remark 7.4.12. If the particle process X is isotropic, we have $\tilde{\mathbb{Q}} \otimes \nu$-almost everywhere $\eta \equiv 1$ by the uniqueness of η in the Definition 7.2.2 of a rotation-regular particle process. Thus, $\overline{\omega}_r(X,t)_p = 0$ for all values of r, t or p.

Remark 7.4.13. The bounds from Remark 7.4.4, which hold for the moduli of smoothness on the rotation group under additional regularity assumptions, imply the following bounds for the modulus of isotropy of first order with $p = 2$. If $\eta(K, \cdot)$ is Lipschitz continuous with constant $L > 0$ for every $K \in \mathcal{K}_0$, i.e.,

$$|\eta(K,\vartheta_1) - \eta(K,\vartheta_2)| \leq L\, d(\vartheta_1,\vartheta_2), \quad \vartheta_1,\vartheta_2 \in SO_d, K \in \mathcal{K}_0,$$

then

$$\overline{\omega}(X,t) \leq t\, L \left(\gamma \mathbb{E} V_d(Z_0 + B^d)\right)^{1/2}.$$

If $\eta(K,\cdot), K \in \mathcal{K}_0$ is uniformly equicontinuous, i.e. there exists for every $\epsilon > 0$ a $\delta(\epsilon) > 0$ such that

$$|\eta(K,\vartheta_1) - \eta(K,\vartheta_2)| \leq \epsilon, \text{ if } d(\vartheta_1,\vartheta_2) \leq \delta(\epsilon), \quad \vartheta_1,\vartheta_2 \in SO_d, K \in \mathcal{K}_0,$$

then

$$\overline{\omega}(X,t) \leq \epsilon \left(\gamma \mathbb{E} V_d(Z_0 + B^d)\right)^{1/2}, \text{ if } t \leq \delta(\epsilon).$$

7.5 Polynomial approximation of Radon-Nikodym derivatives of densities of area measures

Let X be a rotation regular particle process with convex particles. In this section we derive upper bounds for the error in the L^2-norm which arises if the Radon-Nikodym derivatives of the densities of the area measures of X are approximated by spherical polynomials in an optimal way.

Recall from Section 2.4 that π_n is the orthogonal projection on the space $\Pi_n(S^{d-1})$ of spherical polynomials of polynomial degree at most n.

Imposing only an integrability condition we can bound the approximation error of the Radon-Nikodym derivative of they density of an ordinary (i.e. not mixed) area measure in terms of the modulus of isotropy from the previous section.

Lemma 7.5.1. *Let $j \in \{0,\ldots,d-1\}$, $n \in \mathbb{N}_0$, $r \in \mathbb{N}$ and $\overline{\psi}_j(X,\cdot)$ as in Lemma 7.3.1. If $\overline{a}(X) < \infty$, then*

$$\|\overline{\psi}_j(X,\cdot) - \pi_n \overline{\psi}_j(X,\cdot)\|_2^2 \leq c\, \overline{V}_j(X)\, \overline{\omega}_r\left(X, \frac{1}{n+1}\right)^2,$$

where $c > 0$ depends only on d and r.

Proof. Lemma 7.3.1 implies

$$\overline{\psi}_j(X,v) = \gamma \int\limits_{\mathcal{K}_0} \int\limits_{S^{d-1}} \int\limits_{SO_d(e_1)} \eta_{\vartheta\vartheta_u^{-1}K}(v)\, \nu_{e_1}(d\vartheta)\Psi_j(K,du)\tilde{\mathbb{Q}}(dK), \quad v \in S^{d-1}.$$

By the representation Lemma 7.4.9 of the $*$-operator on the kernel η we obtain for each $u \in S^{d-1}$ and $K \in \mathcal{K}_0$ that

$$\int\limits_{SO_d(e_1)} \eta_{\vartheta\vartheta_u^{-1}K}(v)\, \nu_{e_1}(d\vartheta) = \eta^*_{\vartheta_u^{-1}K}(v), \quad v \in S^{d-1}.$$

Thus, we get

$$\overline{\psi}_j(X,v) = \gamma \int\limits_{\mathcal{K}_0} \int\limits_{S^{d-1}} \eta^*_{\vartheta_u^{-1}K}(v)\Psi_j(K,du)\tilde{\mathbb{Q}}(dK), \quad v \in S^{d-1}.$$

Since we can exchange π_n with the double integral in the above equation for $\overline{\psi}_j(X,\cdot)$ by Fubini's theorem we obtain

$$\left[\overline{\psi}_j(X,\cdot) - \pi_n\overline{\psi}_j(X,\cdot)\right](v) = \gamma \int\limits_{\mathcal{K}_0} \int\limits_{S^{d-1}} \left[\eta^*_{\vartheta_u^{-1}K} - \pi_n\eta^*_{\vartheta_u^{-1}K}\right](v)\Psi_j(K,du)\tilde{\mathbb{Q}}(dK). \tag{7.45}$$

For each $K \in \mathcal{K}_0$ and $u \in S^{d-1}$ the bound from Corollary 7.4.10 yields

$$\|\eta^*_{\vartheta_u^{-1}K} - \pi_n\eta^*_{\vartheta_u^{-1}K}\|_2 = \inf_{q\in\Pi_n(S^{d-1})} \|\eta^*_{\vartheta_u^{-1}K} - q\|_2 \le c\,\omega_r\left(\eta(\vartheta_u^{-1}K,\cdot), \frac{1}{n+1}\right)$$

with some $c > 0$ depending only on r and d. The shift property (7.15) of η yields

$$\eta\left(\vartheta_u^{-1}K,\cdot\right) = \eta\left(K,\cdot\vartheta_u^{-1}\right) = T^{\vartheta_u^{-1}}\left[\eta\left(K,\cdot\right)\right].$$

Thus, Lemma 7.4.5 implies

$$\|\eta^*_{\vartheta_u^{-1}K} - \pi_n\eta^*_{\vartheta_u^{-1}K}\|_2 \le c\,\omega_r\left(T^{\vartheta_u^{-1}}\left[\eta(K,\cdot)\right], \frac{1}{n+1}\right) = c\,\omega_r\left(\eta(K,\cdot), \frac{1}{n+1}\right) \tag{7.46}$$

with some $c > 0$ depending only on r and d. Equation (7.45), an application of Hölder's inequality, Fubini's theorem and (7.46) yield

$$\int\limits_{S^{d-1}} \left(\left[\overline{\psi}_j(X,\cdot) - \pi_n\overline{\psi}_j(X,\cdot)\right](v)\right)^2 \sigma(dv)$$

$$= \int\limits_{S^{d-1}} \left(\gamma \int\limits_{\mathcal{K}_0} \int\limits_{S^{d-1}} \left[\eta^*_{\vartheta_u^{-1}K} - \pi_n\eta^*_{\vartheta_u^{-1}K}\right](v)\Psi_j(K,du)\tilde{\mathbb{Q}}(dK)\right)^2 \sigma(dv)$$

$$\leq \gamma \int_{\mathcal{K}_0} \int_{S^{d-1}} \int_{S^{d-1}} \left(\left[\eta^*_{\vartheta_u^{-1}K} - \pi_n \eta^*_{\vartheta_u^{-1}K} \right](v) \right)^2 \sigma(dv) \Psi_j(K, du) \tilde{\mathbb{Q}}(dK) \overline{V}_j(X)$$

$$\leq c\gamma \overline{V}_j(X) \int_{\mathcal{K}_0} \omega_r \left(\eta(K, \cdot), \frac{1}{n+1} \right)^2 V_j(K) \tilde{\mathbb{Q}}(dK).$$

Together with Steiner's formula (2.1) this implies the assertion. □

For the Radon-Nikodym-derivatives of the densities of mixed area measures it is more difficult to obtain upper bounds for the approximation error. The reason is that products of special functions derived from the function η have to be approximated. Thus, we need now the additional assumption that η is bounded on $\mathcal{K}_0 \times SO_d$.

Lemma 7.5.2. *Let* $d \geq 3$, $m = (m_1, m_2) \in \operatorname{mix}(j, 2)$, $j \in \{0, \dots, d-2\}$, $n \in \mathbb{N}_0$ *and* $\overline{\psi}_m(X, X; \cdot)$ *as in Lemma 7.3.1. If* $\|\eta\|_\infty < \infty$, *then*

$$\|\overline{\psi}_m(X, X; \cdot) - \pi_n \overline{\psi}_m(X, X; \cdot)\|_2^2$$
$$\leq c\overline{\omega} \left(X, \frac{1}{n+1} \right)^2 \|\eta\|_\infty^2 \left(\overline{V}_{m_1}(X)^2 \overline{V}_{m_2}(X) + \overline{V}_{m_1}(X) \overline{V}_{m_2}(X)^2 \right),$$

with some $c > 0$ *depending only on d.*

Proof. Lemma 7.3.1 implies

$$\overline{\psi}_m(X, X; \cdot)$$
$$= \gamma^2 \int_{(\mathcal{K}_0)^2} \int_{S^{d-1}} \int_{SO_d(e_1)} \left[\eta_{\vartheta\vartheta_u^{-1}K_1} \eta_{\vartheta\vartheta_u^{-1}K_2} \right](v) \, \nu_{e_1}(d\vartheta) \Psi_m(K_1, K_2; du) \tilde{\mathbb{Q}}^2(d(K_1, K_2)).$$

For $K_1, K_2 \in \mathcal{K}_0$ and $v \in S^{d-1}$ the shift property (7.16) and the definition (7.37) of the $*$-operator imply

$$\int_{SO_d(e_1)} \left[\eta_{\vartheta K_1} \eta_{\vartheta K_2} \right](v) \, \nu_{e_1}(d\vartheta) = \int_{SO_d(e_1)} \left[\eta(K_1, \cdot) \eta(K_2, \cdot) \right]_\vartheta (v) \, \nu_{e_1}(d\vartheta)$$
$$= \left[\eta(K_1, \cdot) \eta(K_2, \cdot) \right]^* (v).$$

Thus, we have

$$\overline{\psi}_m(X, X; \cdot) = \gamma^2 \int_{(\mathcal{K}_0)^2} \int_{S^{d-1}} \left[\eta(\vartheta_u^{-1} K_1, \cdot) \eta(\vartheta_u^{-1} K_2, \cdot) \right]^* (v) \, \Psi_m(K_1, K_2; du) \tilde{\mathbb{Q}}^2(d(K_1, K_2)).$$

By Fubini's theorem the application of π_n can be exchanged with the two-fold integral in the above equation. This yields

$$\int_{S^{d-1}} \left(\left[\overline{\psi}_m(X,X;\cdot) - \pi_n\overline{\psi}_m(X,X;\cdot)\right](v)\right)^2 \sigma(dv)$$
$$= \int_{S^{d-1}} \Big(\gamma^2 \int_{(\mathcal{K}_0)^2}\int_{S^{d-1}} \left(\left[\eta(\vartheta_u^{-1}K_1,\cdot)\eta(\vartheta_u^{-1}K_2,\cdot)\right]^* - \pi_n\left[\eta(\vartheta_u^{-1}K_1,\cdot)\eta(\vartheta_u^{-1}K_2,\cdot)\right]^*\right)(v)$$
$$\times\, \Psi_m(K_1,K_2;du)\tilde{\mathbb{Q}}^2(d(K_1,K_2))\Big)^2 \sigma(dv).$$

Thus, Hölder's inequality and Fubini's theorem imply

$$\|\overline{\psi}_m(X,X;\cdot) - \pi_n\overline{\psi}_m(X,X;\cdot)\|_2^2$$
$$\le \gamma^2 \int_{(\mathcal{K}_0)^2}\int_{S^{d-1}} \left\|\left[\eta(\vartheta_u^{-1}K_1,\cdot)\eta(\vartheta_u^{-1}K_2,\cdot)\right]^* - \pi_n\left[\eta(\vartheta_u^{-1}K_1,\cdot)\eta(\vartheta_u^{-1}K_2,\cdot)\right]^*\right\|_2^2$$
$$\Psi_m(K_1,K_2;du)\tilde{\mathbb{Q}}^2(d(K_1,K_2))\gamma^2 \int_{(\mathcal{K}_0)^2} V_m(K_1,K_2)\tilde{\mathbb{Q}}^2(d(K_1,K_2)). \tag{7.47}$$

For each $K_1, K_2 \in \mathcal{K}_0$ and $u \in S^{d-1}$ the bound from Corollary 7.4.8 for the polynomial approximation of the $*$-operator implies that

$$\left\|\left[\eta(\vartheta_u^{-1}K_1,\cdot)\eta(\vartheta_u^{-1}K_2,\cdot)\right]^* - \pi_n\left[\eta(\vartheta_u^{-1}K_1,\cdot)\eta(\vartheta_u^{-1}K_2,\cdot)\right]^*\right\|_2^2$$
$$\le c\,\omega\left(\eta(\vartheta_u^{-1}K_1,\cdot)\eta(\vartheta_u^{-1}K_2,\cdot), \frac{1}{n+1}\right)^2$$

for some constant $c > 0$ depending only on d. Since Lemma 7.4.5 shows that the map $f \mapsto \omega(f, 1/(n+1))$ is invariant with respect to the operator $T^\varrho, \varrho \in SO_d$, using the bound from Lemma 7.4.6 for the modulus of smoothness of products of functions and applying Hölder's inequality we obtain

$$\omega\left(\eta(\vartheta_u^{-1}K_1,\cdot)\eta(\vartheta_u^{-1}K_2,\cdot), \frac{1}{n+1}\right)^2 = \omega\left(T^{\vartheta_u^{-1}}\left[\eta(K_1,\cdot)\eta(K_2,\cdot)\right], \frac{1}{n+1}\right)^2$$
$$= \omega\left(\eta(K_1,\cdot)\eta(K_2,\cdot), \frac{1}{n+1}\right)^2 \le \left[\omega(\eta(K_1,\cdot)) + \omega(\eta(K_2,\cdot))\right]^2 \|\eta\|_\infty$$
$$\le 2\left[\omega(\eta(K_1,\cdot))^2 + \omega(\eta(K_2,\cdot)^2)\right]\|\eta\|_\infty.$$

Using the above bounds in (7.47), the rotation invariance of $\tilde{\mathbb{Q}}$, the rotational bound (7.20) and Steiner's formula (2.1) we obtain

$$\|\overline{\psi}_m(X,X;\cdot) - \pi_n\overline{\psi}_m(X,X;\cdot)\|_2^2$$

$$
\leq c\,\|\eta\|_\infty^2\,\gamma^2 \int\limits_{(\mathcal{K}_0)^2} \left[\omega\left(\eta(K_1,\cdot),\frac{1}{n+1}\right)^2 + \omega\left(\eta(K_2,\cdot),\frac{1}{n+1}\right)^2\right]
$$

$$
\times V_m(K_1,K_2)\tilde{\mathbb{Q}}(d(K_1,K_2))\gamma^2 \int\limits_{(\mathcal{K}_0)^2} V_m(K_1,K_2)\tilde{\mathbb{Q}}^2(d(K_1,K_2))
$$

$$
\leq c\|\eta\|_\infty^2\,\gamma \int\limits_{\mathcal{K}_0} \omega\left(\eta(K,\cdot),\frac{1}{n+1}\right)^2 V_d(K+B^d)\tilde{\mathbb{Q}}(dK)(\overline{V}_{m_1}(X)+\overline{V}_{m_2}(X))
$$

$$
\times \overline{V}_{m_1}(X)\overline{V}_{m_2}(X)
$$

$$
= c\,\|\eta\|_\infty^2\,\overline{\omega}\left(X,\frac{1}{n+1}\right)^2 (\overline{V}_{m_1}(X)^2\overline{V}_{m_2}(X)+\overline{V}_{m_1}(X)\overline{V}_{m_2}(X)^2)
$$

with some constant $c > 0$ depending only on d. □

In the previous lemmas of this section we were interested in the densities of area measures of the particle process X. Now, we assume that X is in addition a Poisson particle process and Z is the Boolean model, which is the union set of the particles of X. Then, we obtain upper bounds for the L^2-error which arises if the Radon-Nikodym derivatives $\overline{\psi}_j(Z,\cdot)$ of the densities of area measures of the Boolean model with $j \in \{d-1, d-2\}$ are approximated by spherical polynomials.

Lemma 7.5.3. *Let $n \in \mathbb{N}_0$, $r \in \mathbb{N}$ and let X be a rotation regular particle process and Z the associated Boolean model. If $\overline{a}(X) < \infty$, then*

$$
\|\overline{\psi}_{d-1}(Z,\cdot) - \pi_n\,\overline{\psi}_{d-1}(Z,\cdot)\|_2 \leq c\,e^{-\overline{V}_d(X)}\sqrt{\overline{V}_{d-1}(X)}\,\overline{\omega}_r\left(X,\frac{1}{n+1}\right)
$$

with some constant $c > 0$ depending only on d and r.

Proof. Lemma 7.3.2 implies

$$
\overline{\psi}_{d-1}(Z,\cdot) = e^{-\overline{V}_d(X)}\,\overline{\psi}_{d-1}(X,\cdot)
$$

and Lemma 7.5.1 yields

$$
\|\overline{\psi}_{d-1}(X,\cdot) - \pi_n\overline{\psi}_{d-1}(X,\cdot)\|_2 \leq c\sqrt{\overline{V}_{d-1}(X)}\,\overline{\omega}_r\left(X,\frac{1}{n+1}\right)
$$

with some constant $c > 0$ depending only on d and r. Combining both results we obtain the assertion. □

If we approximate the Radon-Nikodym derivative $\overline{\psi}_{d-2}(Z,\cdot)$ by spherical polynomials we obtain the following result.

Lemma 7.5.4. *Let $d \geq 3$, $n \in \mathbb{N}$ and let X be a rotation regular particle process with associated Boolean model Z. If $\|\eta\|_\infty < \infty$, then*

$$\begin{aligned}&\|\overline{\psi}_{d-2}(Z,\cdot) - \pi_n \overline{\psi}_{d-2}(Z,\cdot)\|_2 \\ &\leq c\, e^{-\overline{V}_d(X)} \,\overline{\omega}\left(X, \frac{1}{n+1}\right) \left(\overline{V}_{d-2}(X)^{1/2} + \|\eta\|_\infty \overline{V}_{d-1}(X)^{3/2}\right)\end{aligned}$$

with some constant $c > 0$ depending only on d.

Proof. From Lemma 7.3.2 we get

$$\overline{\psi}_{d-2}(Z,\cdot) = e^{-\overline{V}_d(X)} \left[\overline{\psi}_{d-2}(X,\cdot) - \frac{1}{2}\overline{\psi}_{d-1,d-1}(X,X;\cdot)\right].$$

Lemma 7.5.1 yields

$$\|\overline{\psi}_{d-2}(X,\cdot) - \pi_n \overline{\psi}_{d-2}(X,\cdot)\|_2 \leq c\overline{V}_{d-2}(X)^{1/2}\, \overline{\omega}\left(X, \frac{1}{n+1}\right),$$

with some $c > 0$ which depends only on d. On the other hand Lemma 7.5.2 implies

$$\begin{aligned}&\|\overline{\psi}_{d-1,d-1}(X,X;\cdot) - \pi_n \overline{\psi}_{d-1,d-1}(X,X;\cdot)\|_2 \\ &\leq c\overline{\omega}\left(X, \frac{1}{n+1}\right) \|\eta\|_\infty \overline{V}_{d-1}(X)^{3/2}\end{aligned}$$

with some $c > 0$ depending only on d. The assertion follows by combining both inequalities. □

7.6 A lower bound for the modulus of isotropy

Let X be a rotation regular particle process represented by the triple $(\gamma, \eta, \tilde{\mathbb{Q}})$ with $\overline{a}(X) < \infty$.

For $j \in \{0, \dots, d-1\}$ we denote by

$$\overline{V}_j^n(X) := \left(\overline{V}_j^{n,1}(X), \dots, \overline{V}_j^{n,D(d,n)}(X)\right)$$

the vector of all densities of harmonic intrinsic volumes of degree $n \in \mathbb{N}_0$. Then, the modulus of isotropy

$$\overline{\omega}(X,t) = \left(\gamma \int_{\mathcal{K}_0} \omega(\eta(K,\cdot),t)^2\, V_d(K+B^d)\tilde{\mathbb{Q}}(dK)\right)^{1/2}, \quad t > 0$$

from Definition 7.4.11 with $r = 1$ and $p = 2$ has a lower bound proportional to the Euclidean norm of $\overline{V}_j^n(X)$.

Theorem 7.6.1. *Let $n \in \mathbb{N}$ and $j \in \{0, \dots, d-1\}$, then*

$$\|\overline{V}_j^n(X)\| \leq c\,\overline{V}_j(X)\overline{\omega}\left(X, \frac{1}{n}\right),$$

where $c > 0$ is a constant depending only on d and on the isotropic part $\gamma\tilde{\mathbb{Q}}$ of the distribution of X.

Proof. Observe that by Definition 6.1.1 the densities of the harmonic intrinsic volumes are related to the densities of the area measures by

$$\overline{V}_j^{l,p}(X) = \int_{S^{d-1}} Y_{l,p}(u)\overline{\Psi}_j(X, du).$$

By Lemma 7.3.1 the rotation regularity of X implies that the measure $\overline{\Psi}_j(X, \cdot)$ is absolutely continuous with respect to the spherical Lebesgue measure. Since we have assumed $\overline{a}(X) < \infty$ the bound from Lemma 7.3.3 implies that the Radon-Nikodym derivative $\overline{\psi}_j(X, \cdot)$ is L^2-integrable. Thus, we have

$$\overline{V}_j^{l,p}(X) = (Y_{l,p}, \overline{\psi}_j(X, \cdot)).$$

Therefore, the upper bound for the approximation of $\overline{\psi}_j(X, \cdot)$ by spherical polynomials from Lemma 7.5.1 leads to

$$\begin{aligned}
\|\overline{V}_j^n(X)\|^2 &\leq \sum_{l=n}^{\infty} \sum_{p=1}^{D(d,l)} \overline{V}_j^{l,p}(X)^2 \\
&= \sum_{l=n}^{\infty} \sum_{Y \in B_l} (Y, \overline{\psi}_j(X; \cdot))^2 \\
&= \|\overline{\psi}_j(X; \cdot) - \pi_{n-1}\overline{\psi}_j(X; \cdot)\|_2^2 \\
&\leq c\,\overline{V}_j(X)^2 \overline{\omega}\left(X, \frac{1}{n}\right)^2,
\end{aligned}$$

which implies the assertion. □

Remark 7.6.2. Recall that by the definition of the modulus of isotropy a small value of $\omega\left(X, \frac{1}{n}\right)$ for large n means that X is almost isotropic. The above result could motivate the development of a statistical hypothesis test. For some $t > 0$ define the hypothesis H_0 by

$$H_0: \quad \overline{\omega}\left(X, \frac{1}{n}\right) \leq t.$$

Theorem 7.6.1 suggests that as a test statistic one could use a suitable estimator of the value $\overline{V}_{d-1}(X)^{-1} \|\overline{V}_{d-1}^n(X)\|$.

Remark 7.6.3. If X is a Poisson process, we obtain for $j = d-1$ a lower bound for the modulus of isotropy in terms of the densities of the Boolean model Z related to X. Namely,

$$\overline{V}_{d-1}(Z)^{-1} \left\|\overline{V}^{n}_{d-1}(Z)\right\| \leq c\,\overline{\omega}\left(X, \frac{1}{n}\right),$$

where $c > 0$ is a constant depending only on d.

Remark 7.6.4. Theorem 7.6.1 holds also for the modulus of smoothness of order r with a constant c which depends additionally on r.

7.7 Inversion of density formulas for harmonic intrinsic volumes

As main achievements we obtain in this final section in two and three dimensions a series representation for the intensity of a non-isotropic Boolean model in terms of the densities of the harmonic intrinsic volumes, that is in terms of real-valued observable quantities (see Theorem 7.7.1 in two and Theorem 7.7.6 in three dimensions). For this we combine results from the previous sections of the current chapter. Furthermore, we build on the inversion formulas for the densities of the area measures, which were obtained in the concluding section of Chapter 4. The idea of the proofs is to use the Hilbert space structure on products of the unit sphere. This is possible under the assumption of rotation regularity on the grain distribution, which was introduced in Subsection 7.2, and a suitable integrability condition, which ensures by Subsection 7.3 the absolute continuity of the densities of the area measures with respect to the spherical Lebesgue measure. The disintegration result for the grain distribution from Section 7.1 shows that the assumption of rotation regularity is rather weak. The constants occurring in the series representations are known explicitly as scalar products of spherical harmonic functions on the unit sphere with prescribed functions. As a second step we investigate the error which arises if the intensity is approximated by the truncated series depending on only finitely many densities of harmonic intrinsic volumes. We can bound this error in terms of the modulus of isotropy of the underlying particle process, which was introduced in Section 7.4 (see Theorem 7.7.2 in two and Theorem 7.7.9 in three dimensions). An essential input of the proofs of these approximation results are the results on the polynomial approximation of the Radon-Nikodym derivatives of the densities of the area measures from Section 7.5.

7.7.1 The two dimensional case

Let Z be a rotation regular Boolean model in $\mathbb{R}^2$ represented by the triple $(\gamma, \eta, \tilde{\mathbb{Q}})$ with

$$\overline{a}(X) = \gamma \int_{\mathcal{K}_0} \eta(K, \mathrm{id})^2 V_2\left(K + B^2\right) \tilde{\mathbb{Q}}(dK) < \infty.$$

For $u, v \in S^1$ let $\alpha(u,v) \in [0, \pi]$ be the smaller angle between u and v. Then, we define

$$\varphi(u,v) := |\sin(\alpha(u,v))|\alpha(u,v), \quad u, v \in S^1.$$

Furthermore we use the abbreviation

$$\rho := 1/(1 - \overline{V}_2(Z)) = e^{\overline{V}_2(X)}.$$

In this subsection we write $D(l) := D(2,l), l \in \mathbb{N}_0$ for the dimension of the space of spherical harmonics on S^1 of polynomial degree l. We need the constants

$$c_{l,m}^{p,q} := \pi^{-1}(\varphi, Y_{l,p} \otimes Y_{m,q}), \qquad \text{for } l, m \in \mathbb{N}_0, 1 \leq p \leq D(l), 1 \leq q \leq D(m). \tag{7.48}$$

Then, we can express the intensity by densities of harmonic intrinsic volumes in the following way.

Theorem 7.7.1. *In two dimensions the intensity γ has the series representation*

$$\gamma = \rho \overline{V}_0(Z) + \rho^2 \sum_{l,m=0}^{\infty} \sum_{p=1}^{D(l)} \sum_{q=1}^{D(m)} c_{l,m}^{p,q}\, \overline{V}_1^{l,p}(Z) \overline{V}_1^{m,q}(Z).$$

Proof. The inversion result Corollary 4.2.4 yields the representation

$$\gamma = \overline{V}_0(X) = \rho \overline{V}_0(Z) + \frac{\rho^2}{2} \int_{S^1} \int_{S^1} \mu_2\left(u, v; S^1\right) \overline{\Psi}_1(Z, du) \overline{\Psi}_1(Z, dv), \tag{7.49}$$

where

$$\mu_2\left(u, v; S^1\right) = \frac{2}{\pi} \nabla_2(u,v) \mathcal{H}^1(\mathrm{sconv}\{u, v\}) = \frac{2}{\pi} \varphi(u,v). \tag{7.50}$$

By Lemma 7.3.2 the rotation regularity of X implies that the measure $\overline{\Psi}_1(Z, \cdot)$ is absolutely continuous with respect to the normalized spherical Lebesgue measure with Radon-Nikodym derivative $\overline{\psi}_1(Z, \cdot)$. By Lemma 7.3.4 the integrability condition $\overline{a}(X) < \infty$ is sufficient to ensure that $\overline{\psi}_1(Z, \cdot) \in L^2(S^1)$. Thus, we can use the Hilbert space structure on $L^2((S^1)^2)$ and (7.50) in (7.49) to obtain that

$$\gamma = \rho \overline{V}_0(Z) + \frac{\rho^2}{\pi} (\varphi, \overline{\psi}_1(Z, \cdot) \otimes \overline{\psi}_1(Z, \cdot)). \tag{7.51}$$

An orthonormal basis on $L^2((S^1)^2)$ can be defined by tensor products of orthonormal spherical harmonic functions as in Lemma 2.4.1. Then, the series representation (2.11) of the scalar product implies

$$(\varphi, \overline{\psi}_1(Z,\cdot)\otimes\overline{\psi}_1(Z,\cdot)) = \sum_{l,m=0}^{\infty}\sum_{p=1}^{D(l)}\sum_{q=1}^{D(m)}(\varphi, Y_{l,p}\otimes Y_{m,q})(Y_{l,p}\otimes Y_{m,q}, \overline{\psi}_1(Z,\cdot)\otimes\overline{\psi}_1(Z,\cdot)).$$

Using the Definition 6.1.1 of the harmonic intrinsic volumes it can be shown for each choice of l, m, p and q that

$$\begin{aligned}(Y_{l,p}\otimes Y_{m,q}, \overline{\psi}_1(Z,\cdot)\otimes\overline{\psi}_1(Z,\cdot)) &= (Y_{l,p}, \overline{\psi}_1(Z,\cdot))(Y_{m,q}, \overline{\psi}_1(Z,\cdot))\\ &= \overline{V}_1^{l,p}(Z)\,\overline{V}_1^{m,q}(Z)\end{aligned} \tag{7.52}$$

and thus the assertion. □

In applications only finitely many densities of harmonic intrinsic volumes of the Boolean model can be measured. Thus, we are interested in the truncated series

$$\gamma_n := \rho\,\overline{V}_0(Z) + \rho^2\sum_{l,m=0}^{n}\sum_{p=1}^{D(l)}\sum_{q=1}^{D(m)} c_{l,m}^{p,q}\,\overline{V}_1^{l,p}(Z)\,\overline{V}_1^{m,q}(Z), \quad n\in\mathbb{N}_0. \tag{7.53}$$

Obviously γ_n converges to γ as n goes to infinity. Moreover, it turns out that the speed of convergence is controlled by the modulus of isotropy of the particle process X.

Theorem 7.7.2. *Let $n\in\mathbb{N}_0$. Then*

$$|\gamma-\gamma_n| \le c\,\sqrt{\overline{a}(X)}\,\overline{V}_1(X)\,\overline{\omega}\left(X, \frac{1}{n+1}\right)$$

with some constant $c>0$.

Proof. For $0\le l,m\le n$, $1\le p\le D(l)$ and $1\le q\le D(m)$ we get by the definition (7.48) of the constants $c_{l,m}^{p,q}$ and by the scalar product representation (7.52) of the harmonic intrinsic volumes that

$$\begin{aligned}c_{l,m}^{p,q}\overline{V}_1^{l,p}(Z)\overline{V}_1^{m,q}(Z) &= \frac{1}{\pi}(\varphi, Y_{l,p}\otimes Y_{m,q})(Y_{l,p}\otimes Y_{m,q}, \overline{\psi}_1(Z,\cdot)\otimes\overline{\psi}_1(Z,\cdot))\\ &= \frac{1}{\pi}(\varphi, Y_{l,p}\otimes Y_{m,q})(Y_{l,p}\otimes Y_{m,q}, \pi_n\left(\overline{\psi}_1(Z,\cdot)\otimes\overline{\psi}_1(Z,\cdot)\right)).\end{aligned}$$

Thus, the series representation (2.11) of the scalar product and the definition (7.53) of γ_n imply

$$\gamma_n = \rho\,\overline{V}_0(Z) + \frac{\rho^2}{\pi}\left(\varphi, \pi_n\left[\overline{\psi}_1(Z,\cdot)\otimes\overline{\psi}_1(Z,\cdot)\right]\right).$$

Together with the scalar product representation (7.51) of γ and the Cauchy-Schwarz inequality we get

$$\begin{aligned}|\gamma - \gamma_n| &= \frac{\rho^2}{\pi} \left| \left(\varphi, \overline{\psi}_1(Z,\cdot) \otimes \overline{\psi}_1(Z,\cdot) - \pi_n \left[\overline{\psi}_1(Z,\cdot) \otimes \overline{\psi}_1(Z,\cdot)\right]\right) \right| \\ &\leq \frac{\rho^2}{\pi} \|\varphi\|_2 \, \|\overline{\psi}_1(Z,\cdot) \otimes \overline{\psi}_1(Z,\cdot) - \pi_n \left[\overline{\psi}_1(Z,\cdot) \otimes \overline{\psi}_1(Z,\cdot)\right] \|_2.\end{aligned}$$

Then, we use the upper bound from Lemma 2.4.2 for the L^2-error which arises if tensor products of functions are approximated by spherical polynomials to obtain

$$|\gamma - \gamma_n| \leq \frac{\rho^2}{\pi} \|\varphi\|_2 \, 2 \, \|\overline{\psi}_1(Z,\cdot) - \pi_n \overline{\psi}_1(Z,\cdot)\|_2 \, \|\overline{\psi}_1(Z,\cdot)\|_2.$$

Now, the upper bound from Lemma 7.5.3 for the approximation error which occurs if the Radon-Nikodym derivatives of the densities of the area measures are approximated by spherical polynomials implies

$$\|\overline{\psi}_1(Z,\cdot) - \pi_n \overline{\psi}_1(Z,\cdot)\|_2 \leq c\,\rho^{-1} \sqrt{\overline{V}_1(X)}\, \overline{\omega}\left(X, (n+1)^{-1}\right)$$

with some constant $c > 0$. On the other hand Lemma 7.3.4 yields the bound

$$\|\overline{\psi}_1(Z,\cdot)\|_2 \leq c\,\rho^{-1} \sqrt{\overline{a}(X)\overline{V}_1(X)}$$

with some constant $c > 0$. Thus, we obtain the asserted inequality.

□

Remark 7.7.3. If X is isotropic, the modulus of isotropy is equal to zero and $\gamma = \gamma_0$.

Remark 7.7.4. If $\eta(K,\cdot), K \in \mathcal{K}_0$ is Lipschitz continuous with constant $L > 0$, then Remark 7.4.13 implies that

$$|\gamma - \gamma_n| \leq c\,\frac{1}{n}, \quad n \in \mathbb{N}$$

with some constant $c > 0$ which is independent of n.

Remark 7.7.5. In Theorem 7.7.2 we can replace the modulus of isotropy of first order by the modulus of isotropy $\overline{\omega}_r \left(X, \frac{1}{n+1}\right)$ of order $r \in \mathbb{N}$. Then, the constant c will depend on r.

7.7.2 The three dimensional case

Let Z be a rotation-regular Boolean model in $\mathbb{R}^3$ represented by the triple $(\gamma, \eta, \tilde{\mathbb{Q}})$ with

$$\overline{a}(X) = \gamma \int_{\mathcal{K}_0} \eta(K, \mathrm{id})^2 V_3\left(K + B^3\right) \tilde{\mathbb{Q}}(dK) < \infty.$$

Recall from Section 2.4 that $D(3,l) = 2l+1$ is the dimension of the space of spherical harmonics of polynomial degree $l \in \mathbb{N}_0$ on S^2. We need the functions

$$\varphi_{l,p}(u,v) := \nabla_2(u,v) \int\limits_{\mathrm{sconv}\{u,v\}} Y_{l,p}(w)\mathcal{H}^1(dw), \quad u,v \in S^2, l \in \mathbb{N}_0, 1 \le p \le 2l+1,$$

$$\varphi_3(u,v,w) := \nabla_3(u,v,w)\mathcal{H}^2(\mathrm{sconv}\{u,v,w\}), \quad u,v,w \in S^2,$$

and

$$\xi(u,v) := g(\langle u,v\rangle), \text{ for linearly independent } u,v \in S^2,$$

where

$$g(t) = 1 - t\ln(1+t) - \left(\frac{4}{3} - \ln(2)\right)t, \quad t \in (-1,1).$$

Observe that it holds $\xi \in L^2((S^2)^2)$. This follows since we can bound the L^2-norm of the critical summand in the definition of ξ. Namely, it follows from the rotation invariance of σ and by using polar coordinates that

$$\begin{aligned}
&\int\limits_{S^2}\int\limits_{S^2} (\langle u,v\rangle)^2 \ln(1+\langle u,v\rangle)^2 \sigma^2(d(u,v)) \\
&= \omega_3 \int\limits_{S^2} (\langle e_1,v\rangle)^2 \ln(1+\langle e_1,v\rangle)^2 \sigma(d(v)) \\
&= \omega_3\omega_2 \int\limits_0^\pi (\cos\theta)^2 \ln(1+\cos\theta)^2 \sin\theta d\theta = 8\pi^2\left(\frac{2}{3}\ln(2)^2 + \frac{70}{27} - \frac{16}{9}\ln(2)\right).
\end{aligned}$$

Then, we define for $l,m,o \in \mathbb{N}_0$, $1 \le p \le 2l+1, 1 \le q \le 2m+1$ and $1 \le s \le 2o+1$ the constants

$$c^{p,q,s}_{l,m,o} := \pi^{-1}\left(\varphi_{l,p}, Y_{m,q} \otimes Y_{o,s}\right), \quad d^{p,q}_{l,m} := (\xi, Y_{l,p} \otimes Y_{m,q})$$

and

$$e^{p,q,s}_{l,m,o} := \frac{2}{3\pi}(\varphi_3, Y_{l,p} \otimes Y_{m,q} \otimes Y_{o,s}).$$

We use the abbreviation

$$\rho := 1/(1 - \overline{V}_3(Z)) = e^{\overline{V}_3(X)}.$$

Now, the densities of the harmonic intrinsic volumes $\overline{V}^{l,p}_1(X), l \in \mathbb{N}_0, 1 \le p \le 2l+1$ of the particle process and the intensity γ can be represented in terms of the densities of the harmonic intrinsic volumes of the Boolean model by the following result.

Theorem 7.7.6. *Let $l \in \mathbb{N}_0$ and $1 \le p \le 2l+1$, then*

$$\overline{V}_1^{l,p}(X) = \rho \overline{V}_1^{l,p}(Z) + \rho^2 \sum_{m,o \in \mathbb{N}_0} \sum_{q=1}^{2m+1} \sum_{s=1}^{2o+1} c_{l,m,o}^{p,q,s} \overline{V}_2^{m,q}(Z) \overline{V}_2^{o,s}(Z)$$

and

$$\gamma = \rho \overline{V}_0(Z) + \rho^2 \sum_{l,m \in \mathbb{N}_0} \sum_{p=1}^{2l+1} \sum_{q=1}^{2m+1} d_{l,m}^{p,q} \overline{V}_1^{l,p}(Z) \overline{V}_2^{m,q}(Z)$$
$$+ \rho^3 \sum_{l,m,o \in \mathbb{N}_0} \sum_{p=1}^{2l+1} \sum_{q=1}^{2m+1} \sum_{s=1}^{2o+1} e_{l,m,o}^{p,q,s} \overline{V}_2^{l,p}(Z) \overline{V}_2^{m,q}(Z) \overline{V}_2^{o,s}(Z).$$

Proof. By the Definition 6.1.1 of the harmonic intrinsic volumes we have

$$\overline{V}_1^{l,p}(X) = \int_{S^2} Y_{l,p}(w) \overline{\Psi}_1(X, dw).$$

Thus, the inversion formula for $\overline{\Psi}_1(X, \cdot)$ from Lemma 4.2.1 yields

$$\overline{V}_1^{l,p}(X) = \rho \overline{V}_1^{l,p}(Z) + \frac{\rho^2}{2} \int_{S^2} \int_{S^2} \int_{S^2} Y_{l,p}(w) \mu_2(u, v; dw) \overline{\Psi}_2(Z, du) \overline{\Psi}_2(Z, dv)$$

with the measure

$$\mu_2(u, v; C) = \frac{2}{\pi} \nabla_2(u, v) \mathcal{H}^1(C \cap \operatorname{sconv}\{u, v\}), \quad C \in \mathcal{B}\left(S^2\right).$$

This means

$$\overline{V}_1^{l,p}(X) = \rho \overline{V}_1^{l,p}(Z) + \frac{\rho^2}{\pi} \int_{S^2} \int_{S^2} \varphi_{l,p}(u, v) \overline{\Psi}_2(Z, du) \overline{\Psi}_2(Z, dv).$$

By Lemma 7.3.2 the rotation regularity of X implies that the measure $\overline{\Psi}_2(Z, \cdot)$ is absolutely continuous with respect to the normalized spherical Lebesgue measure with Radon-Nikodym derivative $\overline{\psi}_2(Z, \cdot)$. By Lemma 7.3.4 the integrability condition $\overline{a}(X) < \infty$ yields $\overline{\psi}_2(Z, \cdot) \in L^2(S^2)$. Thus, we can write

$$\overline{V}_1^{l,p}(X) = \rho \overline{V}_1^{l,p}(Z) + \frac{\rho^2}{\pi} (\varphi_{l,p}, \overline{\psi}_2(Z, \cdot) \otimes \overline{\psi}_2(Z, \cdot)). \tag{7.54}$$

By Lemma 2.4.1 the tensor products of orthonormal spherical harmonic functions on the unit sphere form an orthonormal basis of $L^2\left((S^2)^2\right)$. Hence, the series representation (2.11) of the scalar product implies

$$(\varphi_{l,p}, \overline{\psi}_2(Z, \cdot) \otimes \overline{\psi}_2(Z, \cdot))$$

$$= \sum_{m,o\in\mathbb{N}_0} \sum_{q=1}^{2m+1} \sum_{s=1}^{2o+1} (\varphi_{l,p}, Y_{m,q} \otimes Y_{o,s}) \left(Y_{m,q} \otimes Y_{o,s}, \overline{\psi}_2(Z,\cdot) \otimes \overline{\psi}_2(Z,\cdot)\right).$$

For each choice of m, o, q and s the Definition 6.1.1 of the harmonic intrinsic volumes can be used to show that

$$\begin{aligned}(Y_{m,q} \otimes Y_{o,s}, \overline{\psi}_2(Z,\cdot) \otimes \overline{\psi}_2(Z,\cdot)) &= (Y_{m,q}, \overline{\psi}_2(Z,\cdot))(Y_{o,s}, \overline{\psi}_2(Z,\cdot)) \\ &= \overline{V}_2^{m,q}(Z)\,\overline{V}_2^{o,s}(Z).\end{aligned}$$

Using this together with the definition of the constants $c_{l,m,o}^{p,q,s}$ as

$$c_{l,m,o}^{p,q,s} = \frac{1}{\pi}\,(\varphi_{l,p}, Y_{m,q} \otimes Y_{o,s}),$$

in (7.54) yields the first equation of the assertion.

The relation for the intensity can be obtained by similar arguments. Namely, the inversion formula for $\overline{V}_0(X)$ from Corollary 4.2.5 yields

$$\begin{aligned}\gamma = \overline{V}_0(X) = \rho\overline{V}_0(Z) &+ \rho^2 \int_{S^2}\int_{S^2} \xi(u,v)\overline{\Psi}_1(Z,du)\overline{\Psi}_2(Z,dv) \\ &+ \frac{1}{3}\rho^3 \int_{S^2}\int_{S^2}\int_{S^2} \mu_3(u,v,w;S^2)\overline{\Psi}_2(Z,du)\overline{\Psi}_2(Z,dv)\overline{\Psi}_2(Z,dw),\end{aligned}$$

where

$$\mu_3(u,v,w;S^2) = \frac{2}{\pi}\nabla_3(u,v,w)\mathcal{H}^2(\mathrm{sconv}\{u,v,w\}) = \frac{2}{\pi}\varphi_3(u,v,w).$$

Thus, we obtain by the absolute continuity of the densities of the area measures $\overline{\Psi}_1(Z,\cdot)$ and $\overline{\Psi}_2(Z,\cdot)$ with respect to the spherical Lebesgue measure and the L^2-integrability of their Radon-Nikodym derivatives a representation of γ in terms of scalar products as

$$\gamma = \rho\overline{V}_0(Z)+\rho^2\,(\xi, \overline{\psi}_1(Z,\cdot)\otimes\overline{\psi}_2(Z,\cdot))+\frac{2}{3\pi}\rho^3(\varphi_3, \overline{\psi}_2(Z,\cdot)\otimes\overline{\psi}_2(Z,\cdot)\otimes\overline{\psi}_2(Z,\cdot)). \quad (7.55)$$

Now, using the orthonormal basis of $L^2\left((S^2)^2\right)$ consisting of tensor products of spherical harmonic functions from Lemma 2.4.1 and the series representation (2.11) of the scalar product we get

$$(\xi, \overline{\psi}_1(Z,\cdot) \otimes \overline{\psi}_2(Z,\cdot)) = \sum_{l,m\in\mathbb{N}_0} \sum_{p=1}^{2l+1} \sum_{q=1}^{2m+1} (\xi, Y_{l,p} \otimes Y_{m,q}) \left(Y_{l,p} \otimes Y_{m,q}, \overline{\psi}_1(Z,\cdot) \otimes \overline{\psi}_2(Z,\cdot)\right).$$

For each $l, m \in \mathbb{N}_0$, $1 \le p \le 2l+1$ and $1 \le q \le 2m+1$ we have

$$\left(Y_{l,p} \otimes Y_{m,q}, \overline{\psi}_1(Z,\cdot) \otimes \overline{\psi}_2(Z,\cdot)\right) = \left(Y_{l,p}, \overline{\psi}_1(Z,\cdot)\right)\left(Y_{m,q}, \overline{\psi}_2(Z,\cdot)\right)$$

$$= \overline{V}_1^{l,p}(Z)\overline{V}_2^{m,q}(Z).$$

This leads by the definition of the constants

$$d_{l,m}^{p,q} = (\xi, Y_{l,p} \otimes Y_{m,q})$$

to

$$(\xi, \overline{\psi}_1(Z,\cdot) \otimes \overline{\psi}_2(Z,\cdot)) = \sum_{l,m\in\mathbb{N}_0} \sum_{p=1}^{2l+1} \sum_{q=1}^{2m+1} d_{l,m}^{p,q} \overline{V}_1^{l,p}(Z)\overline{V}_2^{m,q}(Z). \tag{7.56}$$

On the other hand we get

$$\begin{aligned}
&(\varphi_3, \overline{\psi}_2(Z,\cdot) \otimes \overline{\psi}_2(Z,\cdot) \otimes \overline{\psi}_2(Z,\cdot)) \\
&= \sum_{l,m,o\in\mathbb{N}_0} \sum_{p=1}^{2l+1} \sum_{q=1}^{2m+1} \sum_{s=1}^{2o+1} (\varphi_3, Y_{l,p} \otimes Y_{m,q} \otimes Y_{o,s}) \\
&\quad \times (Y_{l,p} \otimes Y_{m,q} \otimes Y_{o,s}, \overline{\psi}_2(Z,\cdot) \otimes \overline{\psi}_2(Z,\cdot) \otimes \overline{\psi}_2(Z,\cdot)) \\
&= \sum_{l,m,o\in\mathbb{N}_0} \sum_{p=1}^{2l+1} \sum_{q=1}^{2m+1} \sum_{s=1}^{2o+1} (\varphi_3, Y_{l,p} \otimes Y_{m,q} \otimes Y_{o,s}) \overline{V}_2^{l,p} \overline{V}_2^{m,q} \overline{V}_2^{o,s}.
\end{aligned}$$

Inserting this and (7.56) in (7.55) and by the definition of the constants

$$e_{l,m,o}^{p,q,s} = \frac{2}{3\pi} (\varphi_3, Y_{l,p} \otimes Y_{m,q} \otimes Y_{o,s})$$

we get the assertion. □

Remark 7.7.7. In dimensions $d > 3$ an inversion formula can be obtained by the same arguments for the density $\overline{V}_{d-2}^{l,m}(X)$. For an index smaller than $d-2$ the situation becomes more difficult. For example we cannot express the density $\overline{V}_{d-3}^{l,m}(X)$ by the densities of the harmonic intrinsic volumes of the Boolean model because of the density of the mixed area measure $\overline{\Psi}_{d-2}(Z, u; \cdot)$ which occurs in Lemma 4.2.2.

For an application of the inversion formulas it is necessary that the involved series converge fast enough. Thus, we consider the truncated expressions involving only densities of harmonic intrinsic volumes up to the order $n \in \mathbb{N}_0$. Namely, for $l \in \mathbb{N}_0$ and $1 \le p \le 2l+1$ let

$$\overline{V}_1^{l,p}(X)_n = \rho \overline{V}_1^{l,p}(Z) + \rho^2 \sum_{m,o=0}^{n} \sum_{q=1}^{2m+1} \sum_{s=1}^{2o+1} c_{l,m,o}^{p,q,s} \overline{V}_2^{m,q}(Z)\overline{V}_2^{o,s}(Z) \tag{7.57}$$

and

$$\gamma_n := \rho \overline{V}_0(Z) + \rho^2 \sum_{l,m=0}^{n} \sum_{p=1}^{2l+1} \sum_{q=1}^{2m+1} d_{l,m}^{p,q} \overline{V}_1^{l,p}(Z)\overline{V}_2^{m,q}(Z)$$

$$+\rho^3 \sum_{l,m,o=0}^{n} \sum_{p=1}^{2l+1} \sum_{q=1}^{2m+1} \sum_{s=1}^{2o+1} e_{l,m,o}^{p,q,s} \, \overline{V}_2^{l,p}(Z)\overline{V}_2^{m,q}(Z)\overline{V}_2^{o,s}(Z). \tag{7.58}$$

It is clear that $\overline{V_1}^{l,p}(X)_n$ converges to $\overline{V_1}^{l,p}(X)$ and γ_n to γ as n goes to infinity. In the first case the speed of convergence can be bounded similarly as for the intensity in two dimensions by the modulus of isotropy of the particle process.

Theorem 7.7.8. *Let $n \in \mathbb{N}_0, l \in \mathbb{N}_0$ and $1 \leq p \leq 2l+1$, then*

$$|\overline{V}_1^{l,p}(X) - \overline{V}_1^{l,p}(X)_n| \leq c\,\sqrt{l+1}\,\sqrt{\overline{a}(X)}\,\overline{V}_2(X)\,\overline{\omega}\left(X, \frac{1}{n+1}\right)$$

with some constant $c > 0$.

Proof. For $0 \leq m, o \leq n$, $1 \leq q \leq 2m+1$ and $1 \leq s \leq 2o+1$ the constants $c_{l,m,o}^{p,q,s}$ are defined as

$$c_{l,m,o}^{p,q,s} = \pi^{-1}(\varphi_{l,p}, Y_{m,q} \otimes Y_{o,s})$$

and we have

$$\begin{aligned}
\overline{V}_2^{m,q}(Z)\overline{V}_2^{o,s}(Z) &= (Y_{m,q} \otimes Y_{o,s}, \overline{\psi}_2(Z,\cdot) \otimes \overline{\psi}_2(Z,\cdot)) \\
&= (Y_{m,q} \otimes Y_{o,s}, \pi_n\left[\overline{\psi}_2(Z,\cdot) \otimes \overline{\psi}_2(Z,\cdot)\right]).
\end{aligned}$$

Thus, the definition (7.57) of $\overline{V}_1^{l,p}(X)_n$ and the series representation (2.11) of the scalar product imply

$$\begin{aligned}
\overline{V}_1^{l,p}(X)_n &= \rho\overline{V}_1^{l,p}(Z) + \rho^2 \sum_{m,o=0}^{n} \sum_{q=1}^{2m+1} \sum_{s=1}^{2o+1} \pi^{-1}(\varphi_{l,p}, Y_{m,q} \otimes Y_{o,s}) \\
&\quad \times (Y_{m,q} \otimes Y_{o,s}, \pi_n\left[\overline{\psi}_2(Z,\cdot) \otimes \overline{\psi}_2(Z,\cdot)\right]) \\
&= \rho\,\overline{V}_1^{l,p}(Z) + \rho^2\,\pi^{-1}\,(\varphi_{l,p}, \pi_n\left[\overline{\psi}_2(Z,\cdot) \otimes \overline{\psi}_2(Z,\cdot)\right]).
\end{aligned}$$

Together with (7.54) and the Cauchy-Schwarz inequality this yields

$$\begin{aligned}
&|\overline{V}_1^{l,p}(Z) - \overline{V}_1^{l,p}(Z)_n| \\
&= \rho^2\,\pi^{-1}\,\left|(\varphi_{l,p}, \overline{\psi}_2(Z,\cdot) \otimes \overline{\psi}_2(Z,\cdot) - \pi_n\left[\overline{\psi}_2(Z,\cdot) \otimes \overline{\psi}_2(Z,\cdot)\right])\right| \\
&\leq \rho^2\,\pi^{-1}\,\|\varphi_{l,p}\|_2\,\|\overline{\psi}_2(Z,\cdot) \otimes \overline{\psi}_2(Z,\cdot) - \pi_n\left[\overline{\psi}_2(Z,\cdot) \otimes \overline{\psi}_2(Z,\cdot)\right]\|_2.
\end{aligned}$$

Now the upper bound from Lemma 2.4.2 for the L^2-error which arises if tensor products of functions are approximated by spherical polynomials implies

$$|\overline{V}_1^{l,p}(Z) - \overline{V}_1^{l,p}(Z)_n| \leq \rho^2\,2\pi^{-1}\|\varphi_{l,p}\|_2\,\|\overline{\psi}_2(Z,\cdot) - \pi_n\overline{\psi}_2(Z,\cdot)\|_2\,\|\overline{\psi}_2(Z,\cdot)\|_2.$$

For $u, v \in S^2$ we have

$$|\varphi_{l,p}(u,v)| = \left|\nabla_2(u,v) \int_{\operatorname{sconv}\{u,v\}} Y_{l,p}(w)\mathcal{H}^1(dw)\right| \leq 2\pi \|Y_{l,p}\|_\infty .$$

This implies together with Lemma 2.4.3 and $D(3,l) = 2l+1$ that $\|\varphi_{l,p}\|_2 \leq 2\pi\sqrt{2l+1}$. Furthermore, we obtain by Lemma 7.5.3 that

$$\|\overline{\psi}_2(Z,\cdot) - \pi_n\overline{\psi}_2(Z,\cdot)\|_2 \leq c\,\rho^{-1}\sqrt{\overline{V}_2(X)}\,\overline{\omega}\left(X, \frac{1}{n+1}\right)$$

with some constant $c > 0$. On the other hand Lemma 7.3.4 implies

$$\|\overline{\psi}_2(Z,\cdot)\|_2 \leq c\,\rho^{-1}\sqrt{\overline{a}(X)}\sqrt{\overline{V}_2(X)}$$

with some constant $c > 0$. These bounds imply together the assertion. □

If the intensity γ is approximated by the truncated series γ_n depending only on densities of harmonic intrinsic volumes of order at most n for some $n \in \mathbb{N}_0$, then the approximation error can be bounded from above in terms of the modulus of isotropy of X. The treatment of the intensity in three dimensions is more difficult than in two dimensions. Therefore, we need the additional assumption that the function η is bounded from above on $\mathcal{K}_0 \times SO_n$.

Theorem 7.7.9. *Let $n \in \mathbb{N}_0$ and $\|\eta\|_\infty < \infty$, then*

$$|\gamma - \gamma_n| \leq c\,\overline{\omega}\left(X, \frac{1}{n+1}\right)\overline{b}(X)\,\|\eta\|_\infty$$

with some constant $c > 0$ and

$$\overline{b}(X) := \overline{a}(X)\overline{V}_2(X)^{3/2} + \sqrt{\overline{a}(X)}\,\overline{V}_2(X)^2 + \sqrt{\overline{a}(X)\overline{V}_1(X)\overline{V}_2(X)}.$$

Proof. By equation (7.55) we have

$$\gamma = \rho\overline{V}_0(Z) + \rho^2\left(\xi, \overline{\psi}_1(Z,\cdot)\otimes\overline{\psi}_2(Z,\cdot)\right) + \frac{2}{3\pi}\rho^3\left(\varphi_3, \overline{\psi}_2(Z,\cdot)\otimes\overline{\psi}_2(Z,\cdot)\otimes\overline{\psi}_2(Z,\cdot)\right).$$

On the other hand we have

$$\begin{aligned}\gamma_n = \rho\overline{V}_0(Z) &+ \rho^2 \sum_{l,m=0}^{n}\sum_{p=1}^{2l+1}\sum_{q=1}^{2m+1} d_{l,m}^{p,q}\,\overline{V}_1^{l,p}(Z)\overline{V}_2^{m,q}(Z)\\ &+ \rho^3 \sum_{l,m,o=0}^{n}\sum_{p=1}^{2l+1}\sum_{q=1}^{2m+1}\sum_{s=1}^{2o+1} e_{l,m,o}^{p,q,s}\,\overline{V}_2^{l,p}(Z)\overline{V}_2^{m,q}(Z)\overline{V}_2^{o,s}(Z)\end{aligned}$$

$$= \rho\overline{V}_0(Z) + \rho^2(\xi, \pi_n\left[\overline{\psi}_1(Z,\cdot)\otimes\overline{\psi}_2(Z,\cdot)\right])$$
$$+ \rho^3\frac{2}{3\pi}\left(\varphi_3, \pi_n\left[\overline{\psi}_2(Z,\cdot)\otimes\overline{\psi}_2(Z,\cdot)\otimes\overline{\psi}_2(Z,\cdot)\right]\right).$$

This follows since we have for $0 \le l, m \le n$, $1 \le p \le 2l+1$, $1 \le q \le 2m+1$ the identities

$$d_{l,m}^{p,q} = (\xi, Y_{l,p}\otimes Y_{m,q})$$

and

$$\overline{V}_1^{l,p}(Z)\overline{V}_2^{m,q}(Z) = (Y_{l,p}\otimes Y_{m,q}, \overline{\psi}_1(Z,\cdot)\otimes\overline{\psi}_2(Z,\cdot))$$
$$= \left(Y_{l,p}\otimes Y_{m,q}, \pi_n\left[\overline{\psi}_1(Z,\cdot)\otimes\overline{\psi}_2(Z,\cdot)\right]\right),$$

which implies

$$\sum_{l,m=0}^{n}\sum_{p=1}^{2l+1}\sum_{q=1}^{2m+1} d_{l,m}^{p,q}\,\overline{V}_1^{l,p}(Z)\overline{V}_2^{m,q}(Z) = \left(\xi, \pi_n\left[\overline{\psi}_1(Z,\cdot)\otimes\overline{\psi}_2(Z,\cdot)\right]\right).$$

On the other hand we obtain for $0 \le l, m, o \le n$, $1 \le p \le 2l+1$, $1 \le q \le 2m+1$ and $1 \le s \le 2o+1$ the relations

$$e_{l,m,o}^{p,q,s} = \frac{2}{3\pi}(\varphi_3, Y_{l,p}\otimes Y_{m,q}\otimes Y_{o,s})$$

and

$$\overline{V}_2^{l,p}(Z)\overline{V}_2^{m,q}(Z)\overline{V}_2^{o,s}(Z) = \left(Y_{l,p}\otimes Y_{m,q}\otimes Y_{o,s}, \overline{\psi}_2(Z,\cdot)\otimes\overline{\psi}_2(Z,\cdot)\otimes\overline{\psi}_2(Z,\cdot)\right)$$
$$= \left(Y_{l,p}\otimes Y_{m,q}\otimes Y_{o,s}, \pi_n\left[\overline{\psi}_2(Z,\cdot)\otimes\overline{\psi}_2(Z,\cdot)\otimes\overline{\psi}_2(Z,\cdot)\right]\right),$$

which yield

$$\sum_{l,m,o=0}^{n}\sum_{p=1}^{2l+1}\sum_{q=1}^{2m+1}\sum_{s=1}^{2o+1} e_{l,m,o}^{p,q,s}\,\overline{V}_2^{l,p}(Z)\overline{V}_2^{m,q}(Z)\overline{V}_2^{o,s}(Z)$$
$$= \frac{2}{3\pi}\left(\varphi_3, \pi_n\left[\overline{\psi}_2(Z,\cdot)\otimes\overline{\psi}_2(Z,\cdot)\otimes\overline{\psi}_2(Z,\cdot)\right]\right).$$

Thus, we obtain together with the Cauchy-Schwarz inequality that

$$|\gamma - \gamma_n|$$
$$= \rho^2|(\xi, \overline{\psi}_1(Z,\cdot)\otimes\overline{\psi}_2(Z,\cdot) - \pi_n\left[\overline{\psi}_1(Z,\cdot)\otimes\overline{\psi}_2(Z,\cdot)\right])|$$
$$+ \rho^3\frac{2}{3\pi}|(\varphi_3, \overline{\psi}_2(Z,\cdot)\otimes\overline{\psi}_2(Z,\cdot)\otimes\overline{\psi}_2(Z,\cdot) - \pi_n\left[\overline{\psi}_2(Z,\cdot)\otimes\overline{\psi}_2(Z,\cdot)\otimes\overline{\psi}_2(Z,\cdot)\right])|$$
$$\le \rho^2\|\xi\|_2\|\overline{\psi}_1(Z,\cdot)\otimes\overline{\psi}_2(Z,\cdot) - \pi_n\left[\overline{\psi}_1(Z,\cdot)\otimes\overline{\psi}_2(Z,\cdot)\right]\|_2$$
$$+ \rho^3\frac{2}{3\pi}\|\varphi_3\|_2\|\overline{\psi}_2(Z,\cdot)\otimes\overline{\psi}_2(Z,\cdot)\otimes\overline{\psi}_2(Z,\cdot) - \pi_n\left[\overline{\psi}_2(Z,\cdot)\otimes\overline{\psi}_2(Z,\cdot)\otimes\overline{\psi}_2(Z,\cdot)\right]\|_2.$$

Then, the upper bound from Lemma 2.4.2 for the L^2-error occurring when tensor products of functions are approximated by spherical polynomials yields

$$
\begin{aligned}
&|\gamma - \gamma_n| \\
&\le \rho^2 \|\xi\|_2 \left(\|\overline{\psi}_1(Z,\cdot) - \pi_n \overline{\psi}_1(Z,\cdot)\|_2 \|\overline{\psi}_2(Z,\cdot)\|_2 + \|\overline{\psi}_1(Z,\cdot)\|_2 \|\overline{\psi}_2(Z,\cdot) - \pi_n \overline{\psi}_2(Z,\cdot)\|_2 \right) \\
&\quad + \rho^3 \frac{2}{\pi} \|\varphi_3\|_2 \|\overline{\psi}_2(Z,\cdot) - \pi_n \overline{\psi}_2(Z,\cdot)\|_2 \|\overline{\psi}_2(Z,\cdot)\|_2^2 .
\end{aligned}
$$

From Lemma 7.3.4 we obtain the bounds

$$
\|\overline{\psi}_2(Z,\cdot)\|_2 \le c\rho^{-1} \sqrt{\overline{a}(X) \overline{V}_2(X)}
$$

and

$$
\|\overline{\psi}_1(Z,\cdot)\|_2 \le c\rho^{-1} \sqrt{\overline{a}(X) \overline{V}_1(X) + \frac{1}{2} \overline{a}(X)^2 \overline{V}_2(X)^2}
$$

with some constant $c > 0$. On the other hand Lemma 7.5.3 yields

$$
\|\overline{\psi}_2(Z,\cdot) - \pi_n \overline{\psi}_2(Z,\cdot)\|_2 \le c\rho^{-1} \overline{V}_2(X)^{1/2} \varpi \left(X, \frac{1}{n+1} \right).
$$

Lemma 7.5.4 implies

$$
\begin{aligned}
\|\overline{\psi}_1(Z,\cdot) - \pi_n \overline{\psi}_1(Z,\cdot)\|_2 \le c\rho^{-1} \varpi \left(X, \frac{1}{n+1} \right) \\
\times \left(\overline{V}_1(X)^{1/2} + \|\eta\|_\infty \overline{V}_2(X)^{3/2} \right)
\end{aligned}
$$

with some $c > 0$.

Combining all these bounds we get

$$
\begin{aligned}
&|\gamma - \gamma_n| \\
&\le c \varpi \left(X, \frac{1}{n+1} \right) \|\xi\|_2 \\
&\quad \times \Big[\sqrt{\overline{a}(X) \overline{V}_2(X)} \left(\sqrt{\overline{V}_1(X)} + \|\eta\|_\infty \overline{V}_2(X)^{3/2} \right) \\
&\quad + \left(\sqrt{\overline{a}(X) \overline{V}_1(X)} + \frac{1}{\sqrt{2}} \overline{a}(X) \overline{V}_2(X) \right) \sqrt{\overline{V}_2(X)} \Big] + \frac{2}{\pi} \|\varphi_3\|_2 \, \overline{a}(X) \, \overline{V}_2(X)^{3/2} \\
&\le c \varpi \left(X, \frac{1}{n+1} \right) \|\eta\|_\infty \left[\overline{a}(X) \overline{V}_2(X)^{3/2} + \sqrt{\overline{a}(X)} \, \overline{V}_2(X)^2 + \sqrt{\overline{a}(X) \overline{V}_1(X) \overline{V}_2(X)} \right],
\end{aligned}
$$

with some constant $c > 0$.

□

Remark 7.7.10. If X is isotropic, the modulus of isotropy is equal to zero and

$$\overline{V}_1^{l,p}(X) = \overline{V}_1^{l,p}(X)_0, \quad l \in \mathbb{N}_0, 1 \leq p \leq 2l+1$$

and

$$\gamma = \gamma_0.$$

In other words, in the infinite series in the formulas (7.57) and (7.58) only one summand is not equal to zero.

Remark 7.7.11. If $\eta(K, \cdot)$, $K \in \mathcal{K}_0$ is Lipschitz continuous with constant $L > 0$, then Remark 7.4.13 implies that

$$|\gamma - \gamma_n| \leq c\frac{1}{n} \quad \text{and} \quad \left|\overline{V}_1^{l,p}(X) - \overline{V}_1^{l,p}(X)\right| \leq c\sqrt{l+1}\frac{1}{n}$$

for $l \in \mathbb{N}_0, 1 \leq p \leq 2l+1$ and $n \in \mathbb{N}$ with some constant $c > 0$ which is independent of n.

Remark 7.7.12. In Theorem 7.7.8 the modulus of isotropy of first order can be replaced by the modulus of istropy of order r. Then, the constant c will depend on r.

Bibliography

[Ale96] A. D. Aleksandrov. *Selected Works*. Gordon and Breach, Amsterdam, 1996.

[Ale99a] S. Alesker. Continuous rotation invariant valuations on convex sets. *Ann. of Math. (2)*, 149(3):977–1006, 1999.

[Ale99b] S. Alesker. Description of continuous isometry covariant valuations on convex sets. *Geom. Dedicata*, 74(3):241–248, 1999.

[AKM03] C. Arns, M. Knackstedt, and K. Mecke. Reconstructing complex materials via effective grain shapes. *Phys. Rev. Lett.*, 91(21):1–4, 2003.

[AKM09] C. H. Arns, M. A. Knackstedt, and K. R. Mecke. Boolean reconstructions of complex materials: Integral geometric approach. *Phys. Rev. E*, 80(5):1–17, 2009.

[Art64] E. Artin. *The Gamma Function*. Holt, Rinehart and Winston, New York, 1964.

[AH12] K. Atkinson and W. Han. *Spherical Harmonics and Approximations on the Unit Sphere: An Introduction*. Springer, Berlin, 2012.

[Ber69] C. Berg. Corps convexes et potentiels sphériques. *Danske Vid. Selskab. Mat.-fys. Medd.*, 37(6):1–64, 1969.

[BS91] U. Bindrich and D. Stoyan. Stereology for pores in wheat bread: statistical analyses for the Boolean model by serial sections. *J. Microsc.*, 162(2):231–239, 1991.

[CWZ67] A. P. Calderón, G. Weiss, and A. Zygmund. On the existence of singular integrals. In *Singular Integrals*. American Mathemat. Soc., Providence, 1967.

[Car66] L. Carleson. On convergence and growth of partial sums of Fourier series. *Acta Math.*, 116(1):135–157, 1966.

[CSKM13] S. N. Chiu, D. Stoyan, W. S. Kendall, and J. Mecke. *Stochastic Geometry and its Applications*. Wiley, Chichester, 3. edition, 2013.

[Coh13] D. L. Cohn. *Measure Theory*. Birkhäuser, New York, 2. edition, 2013.

[DX10] F. Dai and Y. Xu. Moduli of smoothness and approximation on the unit sphere and the unit ball. *Adv. Math.*, 224(4):1233–1310, 2010.

[DX13] F. Dai and Y. Xu. *Approximation Theory and Harmonic Analysis on Spheres and Balls*. Springer, New York, 2013.

[Dav76] P. Davy. Projected thick sections through multi-dimensional particle aggregates. *J. Appl. Probab.*, 13(4):714–722, 1976.

[Dit99] Z. Ditzian. A modulus of smoothness on the unit sphere. *J. Anal. Math.*, 79(1):189–200, 1999.

[GL11] D. Gentner and G. Last. Palm pairs and the general mass-transport principle. *Math. Z.*, 267(3-4):695–716, 2011.

[GW02] P. Goodey and W. Weil. Representations of mixed measures with applications to Boolean models. *Rend. Circ. Mat. Palermo, Ser. II Supplemento*, (70):325–346, 2002.

[Had57] H. Hadwiger. *Vorlesungen über Inhalt, Oberfläche und Isoperimetrie*. Springer-Verlag, Berlin, 1957.

[HS71] H. Hadwiger and R. Schneider. Vektorielle Integralgeometrie. *Elem. Math.*, (26):49–57, 1971.

[HHKM14] J. Hörrmann, D. Hug, M. A. Klatt, and K. Mecke. Minkowski tensor density formulas for Boolean models. *Adv. in Appl. Math.*, (55):48–85, 2014.

[HSS08a] D. Hug, R. Schneider, and R. Schuster. Integral geometry of tensor valuations. *Adv. in Appl. Math.*, 41(4):482–509, 2008.

[HSS08b] D. Hug, R. Schneider, and R. Schuster. The space of isometry covariant tensor valuations. *St. Petersburg Math. J.*, 19(1):137–159, 2008.

[Hug99] D. Hug. *Measures, curvatures and currents in convex geometry*. Albert-Ludwig-Universität Freiburg im Breisgau, Freiburg, 1999.

[HS14] D. Hug and R. Schneider. Local Tensor Valuations. *Geom. Funct. Anal.*, 24(5):1516–1564, 2014.

[Kal97] O. Kallenberg. *Foundations of Modern Probability*. Springer, New York, 2. edition, 1997.

[Kal07] O. Kallenberg. Invariant measures and disintegrations with applications to Palm and related kernels. *Probab. Theory Relat. Fields*, 139(1-2):311, 2007.

[Kal11] O. Kallenberg. Invariant Palm and related disintegrations via skew factorization. *Probab. Theory Relat. Fields*, 149(1-2):279–301, 2011.

[KR97] D. A. Klain and G.-C. Rota. *Introduction to Geometric Probability*. Cambridge University Press, Cambridge, 1997.

[Lud02] M. Ludwig. Moment vectors of polytopes. *Rend. Circ. Mat. Palermo (2) Suppl.*, (70):123–138, 2002.

[Lud03] M. Ludwig. Ellipsoids and matrix-valued valuations. *Duke Math. J.*, 119(1):159–188, 2003.

[Lud13] M. Ludwig. Covariance matrices and valuations. *Adv. in Appl. Math.*, 51(3):359–366, 2013.

[Mac09] B. D. MacCluer. *Elementary Functional Analysis*. Springer, New York, 2009.

[Mat75] G. Matheron. *Random Sets and Integral Geometry*. J. Wiley & Sons, New York, 1975.

[McM97] P. McMullen. Isometry covariant valuations on convex bodies. *Rend. Circ. Mat. Palermo, Ser. II Supplemento*, (50):259–271, 1997.

[MS02] K. R. Mecke and A. Seyfried. Strong dependence of percolation thresholds on polydispersity. *Europhys. Lett.*, 58(1):28–34, 2002.

[MW91] K. R. Mecke and H. Wagner. Euler characteristic and related measures for random geometric sets. *J. Stat. Phys.*, 64(3-4):843–850, 1991.

[Mil76] R. E. Miles. Estimating aggregate and overall characteristics from thick sections by transmission microscopy. *J. Microsc.*, 107(3):227–233, 1976.

[Mol97] I. Molchanov. *Statistics of the Boolean Model for Practitioners and Mathematicians*. Wiley, Chichester, 1997.

[Rat97] J. Rataj. The iterated version of a translative integral formula for sets of positive reach. *Rend. Circ. Mat. Palermo, Ser. II Supplemento*, (46):129–238, 1997.

[RG00] A. P. Roberts and E. J. Garboczi. Elastic Properties of Model Porous Ceramics. *J. Am. Ceram. Soc.*, 83(12):3041–3048, 2000.

[Sch72a] R. Schneider. Krümmungsschwerpunkte konvexer Körper (I). *Abh. Math. Semin. Univ. Hambg.*, 37(1):112–132, 1972.

[Sch72b] R. Schneider. Krümmungsschwerpunkte konvexer Körper II. *Abh. Math. Semin. Univ. Hambg.*, 37(3):204–217, 1972.

[Sch78] R. Schneider. Curvature measures of convex bodies. *Ann. Mat. Pura Appl. (4)*, 116(1):101–134, 1978.

[Sch00] R. Schneider. Tensor valuations on convex bodies and integral geometry. *Rend. Circ. Mat. Palermo, Ser. II Supplemento*, (65):295–316, 2000.

[SS02] R. Schneider and R. Schuster. Tensor valuations on convex bodies and integral geometry, II. *Rend. Circ. Mat. Palermo, Ser. II Supplemento*, (70):295–314, 2002.

[Sch13a] R. Schneider. *Convex Bodies: The Brunn-Minkowski Theory.* Cambridge University Press, Cambridge, 2. edition, 2013.

[Sch13b] R. Schneider. Local tensor valuations on convex polytopes. *Monatsh. Math.*, 171(3-4):459–479, 2013.

[SW08] R. Schneider and W. Weil. *Stochastic and Integral Geometry*. Springer, Berlin, 2008.

[SW10] R. Schneider and W. Weil. Classical stochastic geometry. In *New perspectives in stochastic geometry*. Oxford Univ. Press, Oxford, 2010.

[STMK+11] G. E. Schröder-Turk, W. Mickel, S. C. Kapfer, M. A. Klatt, F. M. Schaller, M. J. F. Hoffmann, N. Kleppmann, P. Armstrong, A. Inayat, D. Hug, M. Reichelsdorfer, W. Peukert, W. Schwieger, and K. Mecke. Minkowski tensor shape analysis of cellular, granular and porous structures. *Adv. Mater.*, (23):2535–2553, 2011.

[STMK+13] G. E. Schröder-Turk, W. Mickel, S. C. Kapfer, F. M. Schaller, B. Breidenbach, D. Hug, and K. Mecke. Minkowski tensors of anisotropic spatial structure. *New J. Phys.*, 15(8), 2013.

[Wei90] W. Weil. Iterations of translative integral formulae and non-isotropic Poisson processes of particles. *Math. Z.*, (205):531–549, 1990.

[Wei99] W. Weil. Intensity analysis of Boolean models. *Pattern Recogn.*, 32(9):1675–1684, 1999.

[Wei01a] W. Weil. Densities of mixed volumes for Boolean models. *Adv. in Appl. Probab.*, 33(1):39–60, 2001.

[Wei01b] W. Weil. Mixed measures and functionals of translative integral geometry. *Math. Nachr.*, (223):161–184, 2001.

[Wei07] W. Weil. Random Sets (in particular Boolean models). In *Stochastic Geometry*, volume 1892 of *Lecture Notes in Math.* Springer, Berlin, 2007.

[WW96] E. T. Whittaker and G. N. Watson. *A Course of Modern Analysis*. Cambridge University Press, Cambridge, 4. edition, 1996.